# Propagation and Interference of Spin Waves in Ferromagnetic Thin Films

DISSERTATION ZUR ERLANGUNG DES
DOKTORGRADES DER NATURWISSENSCHAFTEN
(DR. RER. NAT.)
DER NATURWISSENSCHAFTLICHEN FAKULTÄT II -
PHYSIK
DER UNIVERSITÄT REGENSBURG

vorgelegt von

Korbinian Perzlmaier

aus

Mühldorf am Inn

2007

Promotionsgesuch eingereicht am: 10. Oktober 2007

Tag der mündlichen Prüfung: 29. Januar 2008

Die Arbeit wurde angeleitet von: Prof. Dr. Christian H. Back

Prüfungsausschuss:
Vorsitzender: Prof. Dr. John Schliemann
1. Gutachter: Prof. Dr. Christian H. Back
2. Gutachter: Prof. Dr. Dieter Weiss
Prüfer: Prof. Dr. Jascha Repp

Bibliografische Information der Deutschen Nationalbibliothek

Die Deutsche Nationalbibliothek verzeichnet diese Publikation in der Deutschen Nationalbibliografie; detaillierte bibliografische Daten sind im Internet über http://dnb.d-nb.de abrufbar.

ISBN 978-3-8325-1904-9

Logos Verlag Berlin GmbH
Comeniushof, Gubener Str. 47,
10243 Berlin
Tel.: +49 030 42 85 10 90
Fax: +49 030 42 85 10 92
INTERNET: http://www.logos-verlag.de

# Contents

# List of Figures

# Chapter 1

# Introduction

Spin waves have attracted a considerable amount of attention in the last years. First developed by Bloch in 1930 [Blo1930], the concept of spin waves has undergone many changes and refinements [Kit1948, Her1951, Suh1957, Fle1960] and was expanded to such exotic materials as ferromagnetic superconductors [Bra2004]. Since the invention of Magnetic Random Access Memory (MRAM) [Pri1998, Teh1999] several years ago, there has been growing interest in the spin wave resonances of patterned thin magnetic films.

Among these, the first structures to be investigated were field pulse excited magnetic disks, made of Yttrium Iron Garnet (YIG) [Ele1996] or a ferromagnetic metal, observed by means of the Magneto-Optic Kerr Effect (MOKE) [Hie1997, Bue2003, Bue2004-1], Brillouin Light Scattering (BLS) [Jor1999-2, Jor2001, Gub2003], conventional FerroMagnetic Resonance (FMR) [Kak2004], or spatially resolved FMR [Tam2002, Neu2006-2]. The spin wave resonances in these circular structures have also been investigated theoretically [Iva2002] or using micromagnetic simulations [Höl2002, Gio2004], and interesting effects like negative dispersion [Bue2005-1] or mode conversion [Bue2005-2] could be found there, as well as the interaction of spin waves with a magnetic vortex [Par2005, Iva2005, Hof2007]. Other structures investigated include ellipses [Gub2005], rings [Zhu2005-1, Zhu2005-2, Gie2005, Neu2006-1, Gie2007] or stripes [Jor1999-1, Par2002, Bay2003, Cra2003, Bai2003-2, Dem2003-2]. The investigation of stripes, triangular [vKa2006], rectangular [Ger2001, Jor2002, Gus2003, Bay2005-2] or quadratic [Bar2003, Bar2004, Bel2004] structures is of specific interest, as the inhomogeneous magnetic fields inside these structures lead to interesting mode patterns of the magnetic excitation [Par2003, Bay2004, Sto2004, Tam2004, Per2005, Bue2006, Bol2007].

For a long time, the propagation of spin waves has mainly been investigated by BLS in electrically isolating Yttrium Iron Garnet (YIG) films [Esh1962, Ser2004, Wu2006] with a very small damping constant [LeC1958], but it has gained new interest in the last years. Several interesting effects like parametric pumping [Ser2003-1], the formation of solitons [Ser2003-2, Ser2003-3, Wu2004], diffraction [Büt2000], symmetry breaking [Dem2003-1] or the interaction of spin waves with external fields [Dem2004] or with each other [Sla2003], even the formation of a Bose Einstein condensate [Dem2006] have been observed. With the rise of spintronics [Wol2001] and the possibilities to create logic elements based upon spin wave propagation [Esh2007], there was also growing interest in the propagation of spin waves in the metallic ferromagnet $Ni_{80}Fe_{20}$ that can easily be used in lithographic processes and is better suited for miniaturization. Spin waves in $Ni_{80}Fe_{20}$ so far have been observed inductively [Bai2001, Bai2002, Cov2002, Bai2003-1], by FMR [Cou2004] as well as optically [Sil2002, Sch2005, Kos2007, Liu2007] and by simulation [Her2004, Cho2007].

Depending on the geometry of an externally applied magnetic bias field, it is possible to observe negative dispersion [Di1960, Spa1970, Dem2001, Pim2007] for spin waves, an effect that only recently has found growing interest [Pen2003, Par2004] due to interesting consequences like superluminal propagation of wave packets [Chi1993].

Proceeding from the said observations, this thesis deals with the propagation of spin waves in thin $Ni_{80}Fe_{20}$ films. Its specific aims are to experimentally determine a dispersion relation for propagating spin waves in thin ferromagnetic films, and to determine phase velocities, group velocities and damping. Additonally, an analytic dispersion relation based upon the linearized Landau-Lifshitz-Gilbert (LLG) equation is derived, from which the experimental findings can be reproduced using a theoretical model. Further on, this thesis reports for the first time on the direct observation of interference of spin waves, as predicted theoretically [Cho2006, Vas2007]. Up to now, the interference of spin waves has only been realized electrically by converting the spin waves into an electrical microwave signal [Sch2006]. Once the control of spin wave interference has improved, it will be possible to construct interesting elements like interferometers [Sch2006] and more complicated logical elements [Ney2003, Esh2007].

This thesis is structured as follows: In chapter 2, the basic principles of magnetostatics are sketched, followed by the principles of magnetodynamics in chapter 3. Based upon these principles, in chapter 4 a theory for the propagation of spin waves in ferromagnetic thin films is developed.

The experimental part of this thesis begins with an overview of the experimental setup for Time Resolved Scanning Kerr Microscopy (TRSKEM) in chapter 5, and continues with the results of spin wave propagation and interference in chapter 6. The experimental results are always compared to the theoretical predictions of the model developed in chapter 4. Chapter 7 gives a summary of the results and an outlook towards possible future developments. The appendix starting on page 71 gives additional information on experimental and theoretical details.

# Chapter 2

# Theory of Magnetostatics

## 2.1 Magnetic Energies and Fields

Magnetostatics deal with ferromagnetic bodies in static external magnetic fields and with the energies and fields of magnetic bodies. Several energy contributions govern the formation of the magnetic equilibrium configuration; these energies are generally represented by (effective) magnetic fields, linked to the local energy density $\epsilon$ by [Mil2002]

$$\mathbf{H} = -\frac{1}{\mu_0 M_s} \frac{\partial \epsilon}{\partial \mathbf{m}} \quad , \tag{2.1}$$

with the permeability of vacuum $\mu_0$, the saturation magnetization $M_s$, and the direction of the magnetization $\mathbf{m}$. These fields will be treated in the following section. In magnetostatics one mainly deals with the minimization of the overall energy of the system, or, in the picture of magnetic fields, with finding a global magnetization configuration without a net magnetic field. The interactions and dependencies between the different energy contributions lead to the creation of distinct magnetic ground states [Kit1949], hysteresis curves [Sto1948] and many other interesting phenomena such as Bloch-, Néel- and cross-tie walls or magnetic ripples [Fel1960, Fuc1960]. The magnetostatic equations allow calculations of the effective magnetic fields; minimizing these fields leads to the finding of magnetic ground states. When dealing with effective magnetic fields that are not constant in time, one enters the domain of magnetodynamics. As in magnetodynamics commonly only magnetic fields are of interest and not magnetic energies, we shall also stick to magnetic fields and leave aside the equivalent description of magnetic energies.

The information obtained about magnetic fields from a static treatment of the magnetic system can instantly be used with a dynamic approach, as magnetodynamics can merely be treated as a sequence of static magnetic states, combined with damping. Magnetodynamics will be treated in chapter 3. All the formulas in this thesis are given in SI units. The values of the parameters, however, are most of the time given in cgs units, as the cgs system is still very common in magnetism. A conversion table for SI and cgs units is given in appendix D.

## 2.2 Calculation of Magnetic Fields

The effective magnetic field $\mathbf{H}_{\text{eff}}$ is the sum of external magnetic fields, the demagnetizing field which is caused by the magnetization of the sample itself, the exchange field, and material specific crystal anisotropy fields. The latter can be neglected for the case of polycrystalline $Ni_{80}Fe_{20}$ with its very small anisotropy constant. In some special cases, other effects like magnetostriction might become important. However, they do not play a role in the system treated within this thesis.

Throughout this thesis, the magnetization will be labeled as $\mathbf{M}$, with its direction $\mathbf{m} = \mathbf{M}/M_s$, $M_s$ being the saturation magnetization. It is also common to use the magnetic polarization $\mathbf{J}$, defined by $\mathbf{J} = \mu_0\mathbf{M}$. Analogously, $J_s = \mu_0 M_s$ is called the magnetic saturation polarization.

### 2.2.1 External Magnetic Fields

The external magnetic field $\mathbf{H}_{\text{ext}}$ is sometimes referred to as the Zeeman field, as the respective energy $E = \mathbf{H}_{\text{ext}} \cdot \mathbf{M}$ is known as the Zeeman energy. The external magnetic fields in our system can be separated into two components, a static external in-plane bias field $\mathbf{H}_0$ in $x$- or $y$-direction, and high-frequency microwave fields $h_x$ and $h_z$ used for the excitation of magnetization dynamics. In this thesis, these high frequency fields are created by a CoPlanar Stripline (CPS, see section 4.5). Both static and high frequency fields can be treated analogously.

### 2.2.2 Demagnetizing Field

The demagnetizing field is often referred to as the stray field. Starting from Maxwell's equation

$$\nabla \cdot \mathbf{B} = \nabla \left( \mu_0\mathbf{H} + \mathbf{J} \right) = 0 \tag{2.2}$$

the demagnetizing field $\mathbf{H}_d$ is defined as the field generated by the divergence of the magnetic polarization $\mathbf{J} = \mu_0\mathbf{M}$ by

$$\nabla \cdot \mathbf{H}_d = -\nabla \left( \mathbf{J}/\mu_0 \right) . \tag{2.3}$$

10

For explicit calculations, it is useful to switch to the formalism of magnetic charges so that analogies to electrostatics can be used. Within this formalism, magnetic volume and surface charges,

$$\lambda_v = -\nabla \cdot \mathbf{m} \text{ and } \sigma_s = \mathbf{m} \cdot \mathbf{n} \qquad (2.4)$$

(where $\mathbf{n}$ is the surface normal), are introduced. The demagnetizing field potential is then defined by

$$\Phi_{\mathrm{d}}\left(\mathbf{r}\right) = \frac{J_s}{4\pi\mu_0} \left[ \int \frac{\lambda_v\left(\mathbf{r}'\right)}{|\mathbf{r} - \mathbf{r}'|} dV' + \int \frac{\sigma_s\left(\mathbf{r}'\right)}{|\mathbf{r} - \mathbf{r}'|} dS' \right] , \qquad (2.5)$$

and the demagnetizing field can be derived by

$$\mathbf{H}_{\mathrm{d}}\left(\mathbf{r}\right) = -\nabla\Phi_{\mathrm{d}}\left(\mathbf{r}\right) \quad . \qquad (2.6)$$

This means that the well-known formulas from electrostatics can be used to calculate the demagnetizing field of a given magnetization distribution.

The above formulas also imply that the magnetic state of the whole sample has to be known in order to calculate the local demagnetizing field. This fact is presumably the most difficult point in calculating the effective magnetic field. Practically, the demagnetizing field of an arbitrary magnetization distribution can only be calculated numerically. However, in some special cases, an analytic treatment is possible. For the case of spin waves relevant in this thesis, the respective analytic formulas are discussed in section 4.3.

## 2.2.3 Exchange Field

The origin of the exchange field is of purely quantum mechanical nature and cannot be explained in a classical approach. The indistinguishability of electrons and the Pauli principle are the deeper reason for this effect. If two atoms are located near each other, their atomic orbitals overlap. Due to the indistinguishability of the electrons, it is impossible to say which electron is located at which atom. From relativistic quantum mechanics it is known that the combined wave function of the two electrons has to be anti-symmetric for Fermions, i.e. it has to change its sign upon the exchange of the two electrons. It turns out that for ferromagnetic materials, the wave function in position space has a lower energy when it is anti-symmetric, thus the spin wave function has to be symmetric, meaning that the spins of the two electrons have to be aligned parallel. The exact derivation of this is quite complex; a good discussion of this topic can be found in [Whi2007]. The above quantum mechanical effect can be treated as a force,

11

generally referred to as the exchange interaction. Starting from the so-called Heisenberg Hamiltonian

$$\mathcal{H} = \sum_{i \neq j} -A_{ij} \mathbf{S}_i \mathbf{S}_j \tag{2.7}$$

where the $\mathbf{S}_i$ are the electron spins, and $A_{ij}$ is a coupling constant which is positive for ferromagnetic media (and negative for anti-ferromagnetic ones), it is possible to change from the above description of individual spins to a continuum description and to calculate from this Hamiltonian the exchange field by

$$\mathbf{H}_{\text{exch}} = \frac{2A}{\mu_0 M_s^2} \vec{\nabla}^2 \mathbf{M} . \tag{2.8}$$

This field is called local as it can, unlike the demagnetizing field, be calculated locally, without knowing about the overall magnetization configuration.

In effect, as long as the exchange constant $A$ is positive (as it is the case for all ferromagnetic materials), the exchange field imposes an energy penalty to the misalignment of neighboring magnetic moments.

# Chapter 3

# Theory of Magnetodynamics

As mentioned in chapter 2, the magnetic fields calculated with the tools of magnetostatics can also be used in magnetodynamics in order to calculate the dynamic behavior of the magnetization. Magnetic fields can be calculated (at least numerically) for any arbitrary magnetization distribution, and in the Landau-Lifshitz-Gilbert equation (see section 3.1), the effective magnetic fields are linked to dynamic changes of the magnetization via precession and damping.

## 3.1 LLG Equation

The Landau-Lifshitz-Gilbert (LLG) equation is the basic equation of motion for the magnetization. It was first derived by Landau and Lifshitz in 1935 [Lan1935]. The damping term was later modified by Gilbert [Gil1955] in order to remove unphysical results. Gilbert used some previous work by Döring [Dör1948] for his considerations. The LLG equation is given by [Bro1963, Hub2000]

$$\frac{d\mathbf{M}}{dt} = \underbrace{-\mu_0 \gamma \, \mathbf{M} \times \mathbf{H}_{\text{eff}}}_{\text{precession term}} + \underbrace{\frac{\alpha}{M_s} \mathbf{M} \times \frac{d\mathbf{M}}{dt}}_{\text{damping term}} \qquad (3.1)$$

where $\gamma = \frac{ge}{2m_e} = -175.916\frac{\text{GHz}}{\text{T}}$ is the gyromagnetic ratio, $g = 2.0023$ is the $g$-factor of a free electron, $e < 0$ is the elementary charge, $\mu_0 = 4\pi \cdot 10^{-7}\frac{\text{Vs}}{\text{Am}}$ is the permeability of vacuum, and $m_e$ is the mass of the electron. The effective magnetic field used in the above equation is given by

$$\mathbf{H}_{\text{eff}} = \mathbf{H}_{\text{ext}} + \mathbf{H}_{\text{exch}} + \mathbf{H}_{\text{d}} \quad .$$

The LLG equation consists of two terms, a precession term and a damping term. While the precession term describes the precession of the magnetization around

13

its equilibrium position, the damping term provides a torque that pushes the magnetization back to its equilibrium position (see fig.3.1). The direction of the precession and the damping torque are given by the three-fingers-rule of the right hand. It should be mentioned that at the time of its introduction, the damping term was a purely phenomenological description of the observed damping. Today it is known that the damping has a macroscopic origin, and a Gilbert-like term can be predicted from theory [Ho2004-1, Ho2004-2].

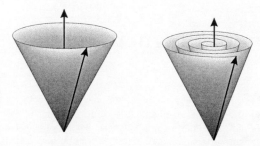

Fig. 3.1: Precession of a magnetic spin around its equilibrium position. Without damping, the spin continuously precesses around its equilibrium position (left), while in a damped system the spin tends to approach its equilibrium position (right).

## 3.2   Spin Wave Theory

The concept of spin waves has been introduced by Bloch in 1930 [Blo1930]. At that time, it was used to explain the $T^{\frac{3}{2}}$ dependence of the saturation magnetization on the temperature. Today, spin waves are a widely spread topic in magnetization dynamics and investigated in quite a number of systems, such as ferrimagnetic **Y**ttrium **I**ron **G**arnet (YIG) films which have a very small damping constant, or ferromagnetic films [Liu2007, Cov2002, Bai2003-1], and their excitation is not only achieved thermally any more, but also by external fields.

Spin waves are collective excitations of spins in a magnetic sample. As with all waves, it is possible to assign a wave vector **k** to the spin wave which points in the direction of its propagation. The absolute value $k$ of the wave vector is referred to as the wave number and related to the wavelength $\lambda$ by $k = \frac{2\pi}{\lambda}$. Mathematically, a spin wave can be expressed by $\mathbf{m}e^{i(\mathbf{k} \cdot \mathbf{x} - \omega t)}$, i.e. a propagating plane wave.

Three different geometries of propagating spin waves in thin films are generally distinguished, which differ in the relative orientation of the wave vector **k** and the magnetization **M**:

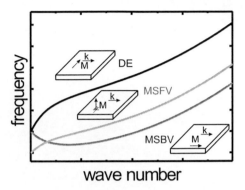

Fig. 3.2: Dispersion relations and configurations of the wave vector **k** and the external magnetic field **M** for the three most common spin wave geometries. Two of these, the DE and the MSBV geometry, have been investigated in this thesis. For higher wave numbers, the frequency of the spin waves increases quadratically due to the exchange field.

- The **D**amon **E**shbach (DE) [Dam1960, Dam1961] or **M**agneto**S**tatic **S**urface **W**ave (MSSW) geometry with the external bias field and the magnetization in the film plane, but perpendicular to the wave vector; the spin waves are mainly located at the surface of the ferromagnetic film.

- The **M**agneto**S**tatic **B**ackward **V**olume (MSBV) geometry with the external bias field and the magnetization in the film plane and parallel to the wave vector; the spin wave amplitude is largest inside the volume of the film. The term "backward" refers to the fact that phase and group velocities have an opposite sign.

- The **M**agneto**S**tatic **F**orward **V**olume (MSFV) geometry with the external bias field and the magnetization oriented perpendicular to the film plane and the wave vector; the spin wave amplitude is largest inside the volume of the film. The term "forward" refers to the fact that phase and group velocities have the same sign.

The three different spin wave geometries and their approximate dispersion relations are sketched in fig. 3.2. There are quite a number of theories to describe the dispersion relations, among them [Hil1990, McM1998, Ari2000]. In this work, an analytic approximation for the dispersions in the DE and MSBV geometries is derived in chapter 4.

# Chapter 4

# Analytic Approximation

The purpose of this chapter is to give an analytic approximation and theoretical explanation for the resonances and dispersion relations observed in the experiments. The aim is to calculate the dynamic susceptibility $\chi$ in the DE and MSBV geometry by linearizing the LLG equation. Then, taking into account the external microwave driving field, the experimentally observed resonances and dispersion relations shall be reproduced.

Within the limits of the **Thin Film Approximation** (TFA, see 4.1), and taking into account the symmetries of the system, it is possible to perform the analytic calculations of the effective magnetic fields, the dynamic susceptibility and the magnetization dynamics in one dimension only (see 4.2). First, the local effective magnetic field for a propagating plane spin wave in the DE and MSBV geometries (see 4.3) will be calculated. With the help of this field, the dynamic susceptibility is derived from the linearized LLG equation (see 4.4). Finally, this result is combined with the microwave excitation spectrum of the CPS to obtain the resonance curves (see 4.6) and dispersion relations (see 4.8).

## 4.1 Thin Film Approximation

The ferromagnetic film is assumed to be thin in the sense of the **Thin Film Approximation** (TFA), a tool widely used to simplify complex magnetic configurations for theoretical treatment. TFA means that the magnetization is assumed to be constant along the thickness of the film, i.e. $\frac{d\mathbf{m}}{dz} = 0$ in our geometry. Nevertheless, the limits of this approximation have to be kept in mind. One might argue that since the DE mode is often described as a surface wave, TFA might not apply to our system; however, TFA is valid as long as the following conditions are met:

- As long as the film thickness has the same magnitude as the exchange length $\lambda_{ex} = \sqrt{\frac{2A}{\mu_0 M_s^2}}$ which is about 6.7 nm for $Ni_{80}Fe_{20}$, the surface effect of the DE spin waves is not too pronounced, as the amplitude of the spin waves cannot decrease notably within the volume of the film.

- The ratio of wavelength to film thickness is also important. When it becomes too small, the coherence between the upper and the lower surface of the film might get lost, and TFA cannot be used anymore.

In summary, for our sample with a thickness of $d = 20$ nm and spin waves with a wavelength of more than 1 $\mu$m, TFA is a valid and good assumption.

## 4.2  Geometric Conventions

We assume a ferromagnetic thin film in the $xy$-plane with a constant thickness $d$ in $z$-direction (see fig. 4.1). A CoPlanar Stripline (CPS) spans over one edge of the ferromagnetic film along the $y$-direction. An external static magnetic in-plane bias field $\mathbf{H}_0$ is applied to the sample, along the $x$-direction for the MSBV geometry and along the $y$-direction for the DE geometry. The CPS creates high frequency microwave fields $h_x$ and $h_z$ around its profile in $x$- and $z$-directions. For the DE and MSBV geometry, the external fields can thus be described by

$$\mathbf{H}_{\text{ext}}^{\text{DE}} = \begin{pmatrix} h_x e^{-i\omega t} \\ H_0 \\ h_z e^{-i\omega t} \end{pmatrix} \text{ and } \mathbf{H}_{\text{ext}}^{\text{MSBV}} = \begin{pmatrix} H_0 + h_x e^{-i\omega t} \\ 0 \\ h_z e^{-i\omega t} \end{pmatrix}, \text{ respectively.}$$

These fields create the dynamic magnetizations

$$\mathbf{M}^{\text{DE}} = \begin{pmatrix} m_x e^{i(kx-\omega t)} \\ M_s \\ m_z e^{i\left(kx-\frac{\pi}{2}-\omega t\right)} \end{pmatrix} \text{ and } \mathbf{M}^{\text{MSBV}} = \begin{pmatrix} M_s \\ m_y e^{i(kx-\omega t)} \\ m_z e^{i\left(kx+\frac{\pi}{2}-\omega t\right)} \end{pmatrix}.$$

The different phases of the two components of the magnetization are due to the precessional motion of the spins, which obeys the three-fingers-rule of the right hand. In both geometries, spin-waves radiate from the CPS in a direction perpendicular to its elongation, i.e. along the $x$-direction, with the wave vector $\mathbf{k} = (k, 0, 0)$ common for both geometries, but with a distinct distribution of wave numbers $k$. Although the wave vectors might be the same for both geometries, it is important to keep in mind that the spin waves described by these wave vectors are of notably different nature.

Due to the symmetry of the problem, none of the parameters depend on $y$, the coordinate pointing along the direction of the CPS. This means that the system has a translational invariance in the $y$-direction. In addition, as within the limits of TFA the magnetization is assumed to be constant in the $z$ direction (see section 4.1), i.e. along the thickness of the film, the problem can be reduced to a one dimensional problem along the $x$ direction, perpendicular to the CPS. These assumptions render the system much less complex and an analytic approach much more promising.

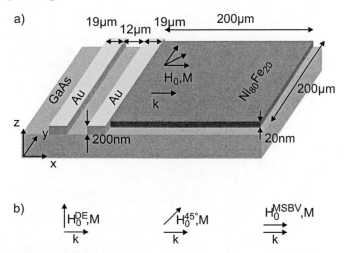

Fig. 4.1: Geometry. a) The CPS is placed over one edge of the $Ni_{80}Fe_{20}$ film, while taking care not to create a shortcut between the two lines of the CPS. b) The three geometries investigated.

## 4.3 Demagnetizing Field Calculation

Several analytic approximations can be used in order to simplify the calculation of the demagnetizing field present in our system. First, an arbitrary magnetization distribution can always be replaced by its Fourier representation (see appendix A.1). This approach is especially promising, as a plane wave is represented by one Fourier component only. Then, it has been stated in section 4.2, that due to certain symmetries, the dynamics of propagating spin waves in thin films can be treated in one dimension only.

However, a purely 1D calculation of the demagnetizing field, as presented in appendix A.2, is too simplified to give the correct results. It indeed delivers an in-plane and a perpendicular demagnetizing field with the correct symmetries expected for the system, but the demagnetizing fields obtained from this purely 1D approach do not depend on the wave number, in contrast to what is expected for real physics. The reason for this is that the purely 1D approach neglects the effect of demagnetizing fields outside the sample.

Thus a more sophisticated approach in calculating the demagnetizing fields of the individual Fourier components has to be chosen. We use a formula given by Harte in [Har1968] which was originally intended to calculate the demagnetizing field of magnetic ripples. Comparing these formulas with the simple 1D calculations, we find that the latter give the correct limits for the case of very high or very low $k$ numbers, depending on the geometry of the demagnetizing field. The formulas by Harte can then be interpreted as a correction to the simple 1D calculations by including the effects of the demagnetizing field outside the sample.

According to [Har1968], the demagnetizing field of an arbitrary magnetization distribution $\mathbf{M}(\mathbf{r}) = \sum_k \mathbf{m_k}(z) e^{i\mathbf{k}\cdot\mathbf{r}}$ can be calculated in the thin film limit $\left(\frac{dm}{dz} = 0\right)$ by[1]

$$(\mathbf{h_k})_m = -\left(\mathbf{k}(\mathbf{k}\cdot\mathbf{m_k})/k^2\right)\tilde{\eta}_k + \hat{\mathbf{e}}_\mathbf{z}(\hat{\mathbf{e}}_\mathbf{z}\cdot\mathbf{m_k})\eta_k \qquad (4.1)$$

with $\eta_k = \frac{1-e^{-kd}}{kd}$, $\tilde{\eta}_k = 1 - \eta_k = \frac{kd-1+e^{-kd}}{kd}$, the wave vector $\mathbf{k} = (k,0,0)$ and the film thickness $d$.

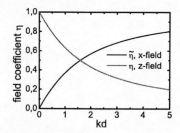

Fig. 4.2: Coefficients $\tilde{\eta}$ for the in-plane, and $\eta$ for the out-of plane demagnetizing field, in dependence on the wave number $k$, used for the calculation of the demagnetizing field of a propagating spin wave in equation 4.1. The coefficients represent the ratio of the demagnetizing field to the magnetization creating this field. As the out-of plane magnetization of a spin wave is generally smaller than its in-plane component due to the elliptic precession in thin ferromagnetic films, the out-of plane demagnetizing fields are substantially smaller than the in-plane fields. The physical meaning of these coefficients is explained in fig. 4.3.

This formula gives a resulting demagnetizing field which is aligned parallel to $\mathbf{k}$ (or the unit vector in $z$-direction, respectively), as expected. This means that due

---

[1]In the following, the letter $\eta$ replaces the letter $\chi$ which was originally used in [Har1968]. This is done to avoid confusion with the susceptibility which is conventionally labeled $\chi$.

to the implied symmetries, only demagnetizing fields perpendicular to the magnetic film, or demagnetizing in-plane fields parallel to **k** appear in the individual components of the Fourier expansion of an arbitrary magnetization configuration. For the calculation of the demagnetizing field, we assume an undamped plane wave with only one contributing Fourier component.

Figure 4.2 shows the coefficients for the demagnetizing field along the $x$- and $z$-directions plotted against the product of wave number $k$ and the film thickness $d$. Figure 4.3 gives a pictorial explanation for the physical reason of this behavior. The complete mathematical derivation of the formula can be found in [Har1968] and shall not be repeated here.

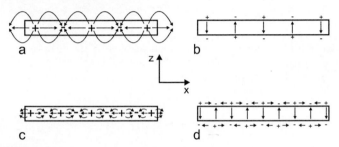

Fig. 4.3: Schematic explanation of the origin of the demagnetizing fields of a spin wave and the physical origin of the coefficients $\bar{\eta}$ and $\eta$.

In-plane demagnetizing field (a,c): Variations of the magnetization along **k** create magnetic volume charges $\lambda = \vec{\nabla} \cdot \mathbf{M}$. Analogous to electrostatics, these charges cause a magnetic field. In a purely 1D calculation as given in appendix A.2, the field is calculated without taking the stray field outside the sample into account and no dependence of the demagnetizing field on the wave number can be found. Taking these fields into account as by equation 4.1, one finds that the purely 1D case corresponds to the high wave number limit (c) of Harte's formula. The coefficient $\bar{\eta}$ thus is a correction term for the stray fields reaching out of the thin film at low wave numbers (a): the longer the distance between the charges (i.e. the smaller the wave number), the more the lines of magnetic flux deviate from a purely in-plane configuration. This leads to an increasing field strength with increasing wave number.

Out of plane demagnetizing field (b,d): Magnetization components along the $z$-axis create magnetic surface charges $\sigma = \mathbf{M} \cdot \mathbf{n}$ when they interact with the sample's surface. These charges in turn create a magnetic field analogous to the one of a plate capacitor (b) which can very easily be calculated in 1D. However, when the distance between alternating charges on one side of the surface becomes smaller, stray fields between these charges on the same side of the surface occur, decreasing the magnetic field inside the magnetic film (d). This leads to a decreasing field strength with increasing wave number. The coefficient $\eta$ can thus be interpreted as a correction term for the creation of stray fields between magnetic charges on the same side of the magnetic film.

## 4.4 Susceptibility Calculations

In this section, we calculate the dynamic susceptibility $\chi$ from the linearized LLG equation for the DE and MSBV geometry. This linearization is only possible for small oscillation amplitudes. For higher precession angles, additional effects would have to be taken into account [Suh1957, Nib2003, Dem2007, Ger2007]. The explicit calculations of these susceptibilities cover several pages and can be found in appendix A.3.

In the following, we consider a continuous magnetic film with a magnetization distribution that can be described by a single plane wave ("spin wave") with a single Fourier component $\mathbf{m}_k$ with the wave number $k$. Due to damping, a realistic spin wave would consist of a distribution of wave numbers. Using only one wave number neglects the damping in the calculation of the demagnetizing field. However, the effect of damping is taken into account by the damping parameter $\alpha$ in the LLG equation 3.1, where it leads to a broadening of the resonance peak and a very small down shift of the resonance frequency, and thus limits the susceptibility to finite values.

In the DE geometry, the external magnetic field

$$\mathbf{H}_{\text{ext}}^{\text{DE}} = \begin{pmatrix} h_x e^{-i\omega t} \\ H_0 \\ h_z e^{-i\omega t} \end{pmatrix} \tag{4.2}$$

leads to the dynamic magnetization

$$\mathbf{M}^{\text{DE}} = \begin{pmatrix} m_x e^{i(kx-\omega t)} \\ M_s \\ m_z e^{i\left(kx-\frac{\pi}{2}-\omega t\right)} \end{pmatrix}. \tag{4.3}$$

According to equation 4.1, the demagnetizing field for the DE geometry is given by

$$\mathbf{H}_{\text{d}}^{\text{DE}} = - \begin{pmatrix} m_x e^{i(kx-\omega t)} \frac{kd-1+e^{-kd}}{kd} \\ 0 \\ m_z e^{i\left(kx-\frac{\pi}{2}-\omega t\right)} \frac{1-e^{-kd}}{kd} \end{pmatrix}. \tag{4.4}$$

Putting this into the LLG equation and linearizing it, the susceptibility can be derived as

$$\chi_{zx}^{\mathrm{DE}} = \frac{m_z}{h_x} = \tag{4.5}$$

$$= \frac{-i\frac{\omega}{\mu_0\gamma}M_s}{\left(\frac{2Ak^2}{\mu_0 M_s} + M_s\frac{1-e^{-kd}}{kd} + H_0 - i\alpha\frac{\omega}{\mu_0\gamma}\right)\left(\frac{2Ak^2}{\mu_0 M_s} + M_s\frac{kd-1+e^{-kd}}{kd} + H_0 - i\alpha\frac{\omega}{\mu_0\gamma}\right) - \frac{\omega^2}{(\mu_0\gamma)^2}} e^{-i\left(kx-\frac{\pi}{2}\right)}$$

for the excitation of spin waves by the $h_x$ field, while the susceptibility for the excitation of spin waves by the $h_z$ field is given by

$$\chi_{zz}^{\mathrm{DE}} = \frac{m_z}{h_z} = \tag{4.6}$$

$$= \frac{-M_s\left(\frac{2Ak^2}{\mu_0 M_s} + M_s\frac{kd-1+e^{-kd}}{kd} + H_0 - i\alpha\frac{\omega}{\mu_0\gamma}\right)}{\left(\frac{2Ak^2}{\mu_0 M_s} + M_s\frac{1-e^{-kd}}{kd} + H_0 - i\alpha\frac{\omega}{\mu_0\gamma}\right)\left(\frac{2Ak^2}{\mu_0 M_s} + M_s\frac{kd-1+e^{-kd}}{kd} + H_0 - i\alpha\frac{\omega}{\mu_0\gamma}\right) - \frac{\omega^2}{(\mu_0\gamma)^2}} e^{-i\left(kx-\frac{\pi}{2}\right)}$$

which has generally the same shape as $\chi_{zx}^{\mathrm{DE}}$, except that its absolute value increases for large $k$ numbers. As $\chi_{zz}^{\mathrm{DE}}$ is at least one order of magnitude smaller than $\chi_{zx}^{\mathrm{DE}}$ for the wave numbers considered in this thesis, it can be neglected for our calculations.

The MSBV case is characterized by a different external magnetic field

$$\mathbf{H}_{\mathrm{ext}}^{\mathrm{MSBV}} = \begin{pmatrix} H_0 + h_x e^{-i\omega t} \\ 0 \\ h_z e^{-i\omega t} \end{pmatrix} \tag{4.7}$$

which leads to the different magnetization configuration

$$\mathbf{M}^{\mathrm{MSBV}} = \begin{pmatrix} M_s \\ m_y e^{i(kx-\omega t)} \\ m_z e^{i\left(kx+\frac{\pi}{2}-\omega t\right)} \end{pmatrix} \tag{4.8}$$

and the demagnetizing field

$$\mathbf{H}_{\mathrm{d}}^{\mathrm{MSBV}} = -\begin{pmatrix} 0 \\ 0 \\ m_z e^{i\left(kx+\frac{\pi}{2}-\omega t\right)}\frac{1-e^{-kd}}{kd} \end{pmatrix}. \tag{4.9}$$

In contrast to the DE geometry, no in-plane demagnetizing field is present to first order. Doing the analogous calculations as in the DE case, the dynamic susceptibility is obtained as

$$\chi_{zz}^{\mathrm{MSBV}} = \frac{m_z}{h_z} = \tag{4.10}$$

$$= \frac{-M_s \left(\frac{2Ak^2}{\mu_0 M_s} + H_0 - i\alpha\frac{\omega}{\mu_0\gamma}\right)}{\left(\frac{2Ak^2}{\mu_0 M_s} + M_s\frac{1-e^{-kd}}{kd} + H_0 - i\alpha\frac{\omega}{\mu_0\gamma}\right)\left(\frac{2Ak^2}{\mu_0 M_s} + H_0 - i\alpha\frac{\omega}{\mu_0\gamma}\right) - \frac{\omega^2}{(\mu_0\gamma)^2}} e^{-i\left(kx+\frac{\pi}{2}\right)}.$$

The absolute values of the susceptibilities are plotted in fig. 4.4. The excitation of spin waves in the MSBV geometry by $h_x$ fields is not possible, thus the respective susceptibility is zero,

$$\chi_{zx}^{\mathrm{MSBV}} = \frac{m_z}{h_x} = 0. \tag{4.11}$$

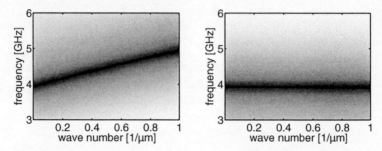

Fig. 4.4: Absolute value of the susceptibilities $\chi_{zx}^{\mathrm{DE}}$ for the DE- (left) and $\chi_{zz}^{\mathrm{MSBV}}$ for the MSBV- (right) geometry. Parameters for the $\mathrm{Ni_{80}Fe_{20}}$ film are (SI units in brackets) $M_s = 860$ Oe ($M_s = 860$ kA/m), $A = 1.05\cdot10^{-6}\ \frac{\mathrm{erg}}{\mathrm{cm}}$ ($1.05\cdot10^{-11}\ \frac{\mathrm{J}}{\mathrm{m}}$), $\alpha = 0.008$, $H_0 = 180$ Oe ($B_0 = 18$ mT, $H_0 \approx 14.3$ kA/m), $d = 20$ nm.

A closer look at the two formulas shows that the two susceptibilities are relations of magnetic fields. This becomes more obvious when they are re-written as

$$\chi_{zx}^{\mathrm{DE}} = \frac{-iH_{gyro}M_s}{\left(H_{\mathrm{exch}} + H_{\mathrm{d,z}} + H_0 - i\alpha H_{gyro}\right)\left(H_{\mathrm{exch}} + H_{\mathrm{d,x}} + H_0 - i\alpha H_{gyro}\right) - H_{gyro}^2} e^{-i\left(kx-\frac{\pi}{2}\right)} \tag{4.12}$$

and

$$\chi_{zz}^{\mathrm{MSBV}} = \frac{-M_s \left(H_{\mathrm{exch}} + H_0 - i\alpha H_{gyro}\right)}{\left(H_{\mathrm{exch}} + H_{\mathrm{d,z}} + H_0 - i\alpha H_{gyro}\right)\left(H_{\mathrm{exch}} + H_0 - i\alpha H_{gyro}\right) - H_{gyro}^2} e^{-i\left(kx+\frac{\pi}{2}\right)}, \tag{4.13}$$

with the fields defined by

$$H_{\text{exch}} = \frac{2Ak^2}{\mu_0 M_s}, \ H_{\text{d,x}} = \frac{M_s\left(kd - 1 + e^{-kd}\right)}{kd}, \ H_{\text{d,z}} = \frac{M_s\left(1 - e^{-kd}\right)}{kd}, \ H_{gyro} = \frac{\omega}{\mu_0\gamma}.$$

As the position of the maximum of the susceptibility is governed by the zeros in the denominator, the denominator mainly governs the shape of the susceptibility. There is one significant difference: In the DE geometry, we find an in-plane and an out-of plane demagnetizing field. In the MSBV geometry, however, only an out-of plane demagnetizing field is found (as there is no in-plane demagnetizing field in the MSBV geometry). Thus the wave number dependence of the maximum frequency in the susceptibilities corresponds to the $k$ dependence of the two demagnetizing fields: increasing with the wave number for the DE geometry, and decreasing with increasing wave number for the MSBV geometry, although less than in the DE geometry.

## 4.5 Wave Number Spectrum of the Driving Microwave Field

The local variation of the driving microwave field created by the CPS can be calculated using standard EM-Simulation software like Empire. From this data, the wave number spectrum of the resulting microwave field can be obtained by means of **Fast Fourier Transformation (FFT)**. Figs. 4.5 and 4.6 show the components $h_x$ and $h_z$ of the driving microwave field and their wave number spectra for the geometries used in the experiments.

An analysis of the magnetic fields created by the CPS shows that the magnetic field in $x$ direction, which is mainly used for the excitation of spin waves in the DE geometry, is strongest underneath the signal line of the CPS. Variations in the strength of the field along the width of the signal line can be explained by the skin effect which leads to increased current densities at the edges of a conductor. Moreover, the current density is stronger at the inner edge than at the outer edge due to the creation of mirror currents in the ground line. Outside the conductor, the magnetic field $h_x$ decreases rather quickly. Due to this field distribution, it is important to place the magnetic film (at least partly) underneath the signal line in order to excite spin waves in the DE geometry, as the field relevant for the excitation of spin waves in the DE geometry is largest here. Fig. 4.5 shows the wave number distribution for the case of a magnetic film below the excitation line, corresponding to the geometry used in continuous and structured leg interference samples (see section 5.4). A wave number distribution with a rather large content of high $k$ values is found here. For another kind of samples, ring interferometers

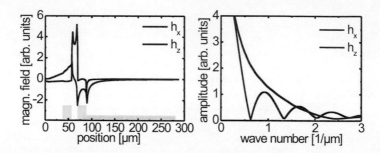

Fig. 4.5: Left: Plot of the high frequency magnetic fields created by the CPS. The position of the CPS is marked by yellow squares, the signal line to the right and the ground line to the left. The film (identical to the region which has been used for the calculation of the wave number spectrum of the microwave driving field) is marked with blue. Its position corresponds to the continuous film sample and the interference sample with legs described in sections 5.4.1 and 5.4.2.1. Due to the skin effect and electrostatic forces, the electric current density and thus the magnetic field is highest at the edges of the signal line, a little higher on the inner edge, while only a small mirror current is found in the ground line with only very weak magnetic fields. The $h_z$ field is strongest between the two conductors, while it is rather small on the outer side of the CPS and decreases quite slowly there. This is the field which can be used for the excitation of spin-waves in the MSBV case. The $h_x$ field, in turn, is strongest underneath the signal line, slightly stronger at the inner edge than at the outer one, due to the different current densities. This is due to the skin effect and the creation of mirror charges on the ground line. The $h_x$ field decreases a lot faster than the $h_z$ field and can be used for the excitation of spin waves in the DE geometry.

Right: wave number spectrum of the microwave driving field created by the CPS, calculated from the fields shown to the left. The exact composition of the wave number spectrum is very sensitive to the position of the edge of the ferromagnetic film underneath the signal line.

The field distribution of the CPS has been calculated for a frequency of 4.0 GHz and does not change notably for the frequency range treated in this thesis.

(see section 5.4.2.2), however, the geometry was different in a way that the film was not placed underneath, but only very close to the CPS (see fig. 4.6). Here, the microwave excitation spectrum contains mostly low wave numbers, which leads to a reduced excitation of high $k$ spin waves in the experiment.

The magnetic field in $z$ direction, in contrast to the $h_x$ field, is strongest in between the two CPS lines, very small underneath them, and decreases outside the conductors, although slower than the $h_x$ field.

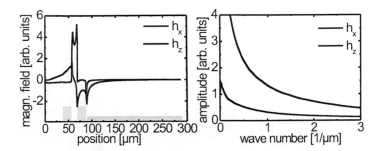

Fig. 4.6: Left: Plot of the high frequency magnetic fields created by the CPS. The position of the CPS is marked by yellow squares, the signal line to the right and the ground line to the left. The film (identical to the region which has been used for the calculation of the wave number spectrum of the microwave driving field) is marked with blue. Its position corresponds to the ring interferometers described in section 5.4.2.2, different to the position of the film in fig. 4.5. Right: wave number spectrum of the microwave driving field created by the CPS, calculated from the fields shown to the left. As the position of the film has slightly changed compared to fig. 4.5, the wave number spectrum has undergone important changes, especially in the $x$ direction. A large decrease in the content of large wave numbers is found.

# 4.6 Resonance Calculations

The amplitude $m_z$ of the resulting oscillation at a fixed frequency $f_0$ can be calculated by summarizing the product of the amplitudes of the wave number components of the microwave driving field $h(k)$ and the absolute value of the respective susceptibility $\chi(f_0, k)$:

$$m_z\left(f_0^{\mathrm{DE}}\right) = \sum_k h_x(k) \left|\chi_{zx}^{\mathrm{DE}}(f_0, k)\right| \tag{4.14}$$

or

$$m_z\left(f_0^{\mathrm{MSBV}}\right) = \sum_k h_z(k) \left|\chi_{zz}^{\mathrm{MSBV}}(f_0, k)\right| \tag{4.15}$$

Plotting the amplitudes against the frequency, the resonance frequencies can be determined easily (see fig. 6.1). These formulas imply that the observed resonance frequency depends on the wave number spectrum of the driving microwave field, an effect observed in [Sch2004].

The above formulas state that for the calculation of the resonance frequency, the values of constant frequency in the susceptibility (i.e. horizontal lines in fig. 4.4) have to be added up and weighted with the proper factors given by the excitation spectrum. Essentially, this is equivalent to the standard product of a vector and a matrix, given by $\vec{m} = |\chi|\,\vec{h}$.

## 4.7 Wavelength Calculations

In standard dispersion relations, the frequency and the wavelength of an oscillation in resonance conditions are directly linked to each other. However, in our case, it is possible to excite the magnetization at frequencies and wave numbers which are far away from the resonance conditions. While the frequency can be chosen freely with an increment of 80 MHz, the respective wave number distribution of this excitation is given by the CPS geometry. Moreover, the driving microwave field does not only contain one discrete wave number $k$, but a complete spectrum of wave numbers (see figs. 4.5 and 4.6). The observed wave number will thus be governed by the susceptibility and the wave number spectrum of the microwave driving field.

In the DE geometry, the $h_x$ field is mainly responsible for the excitation of spin waves, as $\chi_{zz}^{\text{DE}}$ is more than one order of magnitude smaller than $\chi_{zx}^{\text{DE}}$ for the wave numbers of interest. In the MSBV geometry, however, only the $h_z$ field can excite spin waves, as $\chi_{zx}^{\text{MSBV}}$ is zero.

The wave number spectrum for an oscillation at a fixed excitation frequency $f_0$ can be calculated by

$$m_z^{\text{DE}}(k)\Big|_{f_0} = \chi_{zx}^{\text{DE}}(k, f_0)\, h_x(k) \qquad (4.16)$$

for the DE and

$$m_z^{\text{MSBV}}(k)\Big|_{f_0} = \chi_{zz}^{\text{MSBV}}(k, f_0)\, h_z(k) \qquad (4.17)$$

for the MSBV geometry. The above formulas state that in order to calculate the resonant wavelength of a spin wave, the susceptibility at a fixed frequency (i.e. horizontal lines in fig. 4.4) has to be weighted with the proper factors given by the wave number spectrum of the microwave driving field. The peak of the obtained spectrum then represents the resonant wave number. The wave numbers observed by TRSKEM correspond to these peaks. As the numerically obtained wave number spectrum of the microwave field is taken into account for these calculations, no analytic value for the obtained wave lengths can be given. In the DE geometry, an increase of the wave number with increasing frequency is observed because it is very easy to obtain a peak at high $k$ values in this case, as for a constant frequency $f_0$ the susceptibility has a peak for $k \neq 0$. In the MSBV geometry, in contrast, the slope of the maximum in the susceptibility is so small that for the given wave number distribution of the driving field, no isolated maximum can be expected for nonzero $k$ numbers in the wave number spectrum. Thus the (small) negative dispersion in the MSBV geometry cannot be observed.

The wave numbers observed by TRSKEM correspond to the peaks in the above spectrum (see fig 4.7), wave number $k$ and wavelength $\lambda$ are linked to each other by $\lambda = \frac{2\pi}{k}$. Evaluating the the peaks in this wave number distribution, it is possible to derive a dispersion relation from the data (see section 4.8).

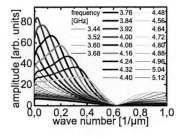 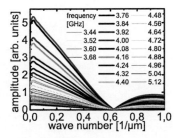

Fig. 4.7: Wave number plots for the DE (left) and the MSBV (right) geometry, calculated with formulas 4.16 and 4.17 for the continuous film sample described in section 5.4.1. The peaks in the wave number spectrum correspond to the wavelengths obtained experimentally upon excitation with a fixed frequency. The amplitudes of the two geometries can directly be compared to each other: This means that the oscillation amplitudes in the DE geometry are by a factor of approximately 10 larger than in the MSBV geometry, owing to the larger susceptibility in the DE geometry.

Comparing the two wavelength spectra, some clearly identifiable peaks at different wave numbers for different excitation frequencies $f_0$ are found in the DE geometry, while in the MSBV geometry the wave number maximum is close to zero for all frequencies. This is due to the shallow susceptibility in the MSBV geometry, which makes it impossible to create nonzero resonant wave numbers for the given excitation spectrum. In the DE geometry, in contrast, the rather steep susceptibility allows for the creation of various spin waves with nonzero wave numbers.

## 4.8 Dispersion Relation Calculations

A dispersion relation for the observed spin waves can be obtained directly from the wavelengths calculated in section 4.7. There exists a second possibility to calculate the dispersion analytically, without taking into account the excitation spectrum: The denominator of the susceptibility can be set to zero, and the obtained equation can be solved for $k$ using standard mathematical toolboxes, such as Maple. The obtained equation for the DE geometry is (with $\gamma < 0$)

$$f_{\text{res}}^{\text{DE}} = \frac{\mu_0 \gamma}{4\pi (\alpha^2 + 1)} \cdot \tag{4.18}$$

$$\cdot \left[ \sqrt{4H_{\text{d,x}} H_{\text{d,z}} + 4 (H_{\text{exch}} + H_0)(H_{\text{exch}} + H_{\text{d,x}} + H_{\text{d,z}} + H_0) - \alpha^2 (H_{\text{d,x}} - H_{\text{d,z}})^2} + \right.$$

$$\left. + i\alpha (2H_{\text{exch}} + H_{\text{d,x}} + H_{\text{d,z}} + 2H_0) \right],$$

and for the MSBV geometry

$$f_{\text{res}}^{\text{MSBV}} = -\frac{\mu_0 \gamma}{4\pi (\alpha^2 + 1)} \cdot \tag{4.19}$$

$$\cdot \left[ \sqrt{4 (H_{\text{exch}} + H_0)(H_{\text{exch}} + H_{\text{d,z}} + H_0) - \alpha^2 H_{\text{d,z}}^2} + \right.$$

$$\left. + i\alpha (2H_{\text{exch}} + H_{\text{d,z}} + 2H_0) \right],$$

where the magnetic fields are defined by

$$H_{\text{exch}} = \frac{2Ak^2}{\mu_0 M_s}, \; H_{\text{d,x}} = \frac{M_s \left( kd - 1 + e^{-kd} \right)}{kd}, \; H_{\text{d,z}} = \frac{M_s \left( 1 - e^{-kd} \right)}{kd}.$$

The only difference in these formulas is that for the MSBV geometry no demagnetizing field in the $x$ direction exists. The obtained (complex) result delivers the frequency of maximum susceptibility, what could be called the classical susceptibility. The resonance frequency is given by the absolute value of these complex frequencies, while the damping is given by its imaginary parts (see section 4.9).

It is worth mentioning that these analytic dispersions differ slightly from the results obtained in section 4.7, as those include the effect of the wave number distribution of the microwave driving field. Results agree well with the experimental observations (see 6.2.3). The analytic dispersion relations are plotted in fig. 4.8 for the DE and MSBV geometries.

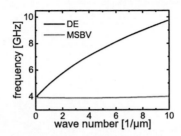

Fig. 4.8: Analytic dispersion relations for Ni$_{80}$Fe$_{20}$ material parameters (same as in fig. 4.4) and an external bias field $H_0 = 180$ Oe (18 mT) in the DE and MSBV geometries.

## 4.9  Damping Time Calculations

The damping time is related to the imaginary part of the resonance frequency (equations 4.18 and 4.19) by

$$\tau^{\mathrm{DE}} = \frac{1}{2\pi \mathrm{Im}\left(f_{\mathrm{res}}^{\mathrm{DE}}\right)} = \frac{-2\left(\alpha^2 + 1\right)}{\alpha\mu_0\gamma\left(2H_{\mathrm{exch}} + H_{\mathrm{d,x}} + H_{\mathrm{d,z}} + 2H_0\right)} \qquad (4.20)$$

for the DE, and

$$\tau^{\mathrm{MSBV}} = \frac{1}{2\pi \mathrm{Im}\left(f_{\mathrm{res}}^{\mathrm{MSBV}}\right)} = \frac{-2\left(\alpha^2 + 1\right)}{\alpha\mu_0\gamma\left(2H_{\mathrm{exch}} + H_{\mathrm{d,z}} + 2H_0\right)} \qquad (4.21)$$

for the MSBV geometry.

The damping time is only slightly dependent on the wave numbers and frequencies as long as they are in the range considered in this thesis and can be calculated to $\tau \approx 1.3$ ns.

# Chapter 5

# Experimental Setup

We use a very sophisticated setup of **T**ime **R**esolved **S**canning **K**err Microscopy (TRSKEM) which has proven its fine capabilities in the observation of magnetization dynamics over the last years [Acr2000, Bue2003, Bue2004-2, Bue2005-3]. The basis of this stroboscopic experiment is given by the light source, a Ti:sapphire femtosecond pulsed laser and a MOKE (**M**agneto **O**ptic **K**err **E**ffect) detector, to which several different types of magnetization excitation can be synchronized, such as magnetic pulses [Bue2004-1, Bue2005-1, Per2005], laser pulses [Acr2001] or continuous microwaves [Neu2006-2]. For the experiments covered in this thesis, two different types of microwave excitation, continuous **w**ave (cw) excitation and **M**icro**W**ave **P**ulse (MWP) excitation are used. As the laser and the microwave parts of the setup can be changed mostly independently from each other and are only linked to each other by the pulse synchronization, the two systems shall be explained independently in the following sections, starting with the laser system in section 5.1, continuing with the two different microwave setups in section 5.2 and the electronic data acquisition in section 5.3, and concluding with the sample layout in section 5.4.

## 5.1 Laser System

The basis of the experimental setup, the light source, is a Coherent Mira 900 femtosecond Ti:sapphire pulsed laser with a repetition rate of 80 MHz synchronized to an external reference frequency and a pulse length of about 200 fs. Short laser pulses at a fixed repetition rate thus allow for the observation of magnetization dynamics with a time resolution better than 1 ps in a stroboscopic experiment. The Mira 900 Ti:sapphire pulsed laser is pumped by a Coherent Verdi V-10 Nd:vanadate cw laser, which, in turn, is pumped by a diode laser array.

The laser beam from the Ti:sapphire oscillator is originally infrared, with a wavelength of about 800 nm and a power of 1 W, and guided through a telescope for

33

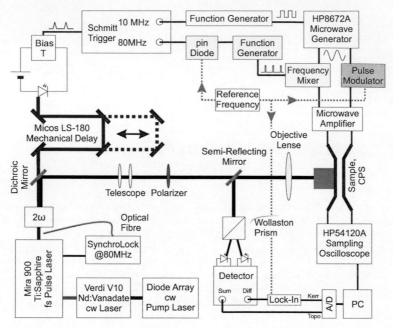

Fig. 5.1: Schematic sketch of the experimental setup, consisting of the following elements:

– Laser system, for details see section 5.1:
• Pulsed laser source with pump lasers and frequency doubler (bottom left)
• Red laser beam used to synchronize microwave excitation, and mechanical delay stage (top left)
• Blue laser with telescope, polarizer and objective lens as used for the observation of the sample's magnetization (center row)

– Microwave system, for details see section 5.2:
• Phase stable synchronization to laser repetition frequency of 80 MHz (top left)
• Creation of the microwave excitation with elements used for cw (light violet) and MWP (light yellow) excitation only (top right)
• Sample and observation of microwaves in transmission (bottom right)

– Data acquisition with detector and lock-in amplifier (bottom right, details see section 5.3)

divergence control right behind the exit slit. As the lateral resolution of a microscope is given by the diffraction limit defined by $d = \frac{0.61\lambda}{N_a}$ (with $N_a$ being the numerical aperture of the used objective), the laser beam is sent through a lithium borate crystal ($LiB_3O_5$, LBO) which by Second Harmonic Generation (SHG) converts a part of the 800 nm beam to 400 nm blue light. Due to the smaller wavelength, the optical resolution is improved to ≈250 nm for the used objective with a numerical aperture of 0.85. The two beams are subsequently separated from each other by a dichroic beam splitter.

The red beam is then guided through a mechanical delay line, after which a part of the red laser beam is sent into a photo diode to create the synchronization signal for the microwave source (see section 5.2), while the remainder of the red laser beam is dumped. The mechanical delay line allows to change the delay between the blue laser probe pulses and the microwave excitation synchronized to the red laser pulses. It is possible to scan the phase between microwave excitation and the blue laser probe pulses. The distance $\Delta x$ and delay $\Delta t$ are linked to each other by the speed of light. A delay of 1 ns corresponds to a distance of 30 cm.

The blue laser beam, in contrast, is guided through a telescope for divergence control and then fed through a linear polarizer and into a Zeiss Axiomat microscope. Inside the microscope, the blue laser beam is sent through the objective lens and reflected from the sample. The reflected beam has a slightly rotated polarization due to the magneto-optic Kerr effect (see appendix B). As perpendicular incidence is used, the setup is sensitive to the out-of plane component of the magnetization. The reflected beam is guided back through the same objective lens and subsequently through a Wollaston prism which splits the reflected beam into two beams with directions of polarization perpendicular to each other.

The two beams then illuminate two photo diodes of an optical bridge in the detector. The sum and the difference of the signal of the two diodes are detected and pre-amplified. The sum signal is proportional to the reflectivity of the sample and often referred to as the topography signal. The difference signal, in turn, is proportional to the out-of plane magnetization of the sample and referred to as the Kerr signal. It is worth to point out that the photo diodes have a low bandwidth of 10 kHz and cannot follow the 80 MHz repetition rate of the laser. The bandwidth is however large enough to easily follow the chopping frequency of 1-3 kHz of the microwave excitation (see section 5.2). The following steps of detection and modifying the electric signal are described in section 5.3.

## 5.2 Microwave System

One of the biggest challenges of the complete setup is to establish a phase stable synchronization between the blue laser probe pulses and the microwave excitation in order to be able to conduct a stroboscopic experiment.

The red laser beam is sent through a mechanical delay line (see section 5.1) and then onto a fast photo diode attached to a bias voltage. The photo diode has a rise time of 50 ps. Using a bias-T, this analog high frequency response of the photo diode is separated from the bias voltage line and sent to a Schmitt trigger

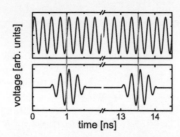

Fig. 5.2: Schematic of the synchronization of laser pulses and microwave excitation, for a cw (above) and MWP excitation (below). The pulse laser with a repetition rate of 80 MHz delivers a flash of light each 12.5 ns (grey lines). This means that the phase of the excitation has to be the same each 12.5 ns, too.

which turns it into an ECL signal. In addition to the 80 MHz digital output of the Schmitt trigger, the frequency of this signal is also divided by 8 (i.e. modified to a frequency of 10 MHz) and sent to the external trigger input of a HP8082A function generator that generates a square wave signal as reference frequency for the HP8672A microwave generator. The HP8672A microwave generator is a tunable microwave source whose frequency should be changed in steps of multiples of 80 MHz, the repetition rate of the laser, in order to keep the phase between excitation and observation stable (see fig. 5.2).

The output of the microwave generator is then used as the excitation signal for the magnetization of the sample. However, the details of how this is achieved vary for the cw and MWP type of excitation. The two cases are treated independently in sections 5.2.1 and 5.2.2. This microwave excitation is then applied to the sample.

After the excitation signal has passed through the sample, it is recorded by a HP54120A sampling oscilloscope. This is done in order to control the phase stability, and the correct shaping of the microwave pulse in the MWP excitation case.

## 5.2.1 Continuous Wave Excitation

The output of the microwave generator is guided through a HP11720A pulse modulator attached to a reference frequency generator of 1-3 kHz. This pulse modulator switches the microwaves on and off in order to allow lock-in measurements. Subsequently, the chopped microwaves are amplified by a microwave amplifier and sent to the sample, where they are used for the excitation of the spin wave dynamics.

## 5.2.2 Microwave Pulse Excitation

The creation of microwave pulses for the excitation is somewhat more complex. In addition to the output of the HP8672A microwave generator, an additional high frequency pulse is needed, which defines the length of the the microwave pulse. For that purpose, a HP8131A function generator is used. The external trigger input of the pulse generator is supplied by the 80 MHz output of a Schmitt trigger (see section 5.2) to create a rectangular pulse with a repetition rate of 80 MHz and a pulse length of 1 ns. This pulse signal is sent to a frequency mixer, together with the continuous microwave signal. The frequency mixer multiplies the two input signals and delivers a microwave pulse. It is worth mentioning that the ring mixers do not work ideally, i.e. they suppress the signal by ≈20 dB only when the pulse signal is switched off. As a result, small continuous microwave oscillations are still present besides the actually desired microwave pulse.

However, this effect can be minimized by applying a different chopping scheme (the frequency mixer in this setup only shapes the microwave pulses, and does not chop the signal). In the cw excitation case, the continuous microwave signal is chopped by a pulse shaper which suppresses the signal by ≈80 dB. It would be possible to insert the chopper after the ring mixer and use lock-in measurements. However, this is of no use when it comes to optimizing the bad signal suppression of the frequency mixer. A better idea is to insert a pin diode (HP11665B) into the trigger signal of the pulse generator used for the creation of the 1 ns rectangular pulse and to chop the creation of the microwave pulses.

Using the setup described above, not the complete microwave excitation is chopped, but the creation of microwave pulses by the frequency mixer is turned on and off with the reference frequency of 1-3 kHz. As lock-in measurements always deliver the difference between 'on' and 'off' as a result, the difference between MWP excitation and suppressed cw excitation is measured by this chopping scheme. This notably decreases leakage effects from the frequency mixer.

After the pulse is shaped by the frequency mixer, it is amplified by broadband microwave amplifiers and sent to the sample where it is used to excite spin wave packets. With the used amplifiers, microwave powers of up to 30 dBm can be reached.

# 5.3 Data Acquisition

As the reflectivity of the sample is not changed by magnetization dynamics, the topography signal is a DC signal and can be detected with an AD (**A**nalog/**D**igital) converter card directly. The Kerr signal, in contrast, which is proportional to the out-of plane component of the magnetization, is an AC signal which oscillates with the chopping frequency of the microwave excitation, i.e. the reference frequency of 1-3 kHz. The reason for this is that the laser beam probes the ground state of the magnetization when the excitation is switched off, and the excited magnetization state at a specific time after the excitation when the microwave excitation is switched on. Using lock-in amplification to increase the signal/noise ratio, the lock-in amplifier then delivers a DC signal which is proportional to the difference of the Kerr signal of the excited state and that of the not excited state. This DC Kerr signal is as well detected by an AD converter card in a personal computer which is controlled by a LabView program.

## 5.4   Sample Layout

Two different types of samples have been investigated in this work. The first sample is a (semi) continuous $Ni_{80}Fe_{20}$ film (see section 5.4.1) which has been used for the observation of the propagation of spin waves and spin wave packets. Then two kinds of structured thin film $Ni_{80}Fe_{20}$ samples have been used, a continuous film with leg structures (see section 5.4.2.1), and a kind of ring interferometer structures (see section 5.4.2.2) which have also been used for the detection of spin wave interference. All the samples have been produced in Prof. Weiss' clean room facility.

In the microscope, the sample is mounted on a piezo scanning table which allows to scan the sample up to 80 $\mu$m in both lateral directions. A four-quadrant-magnet allows the application of a static external magnetic in-plane bias field to the sample in different geometries (see fig. 4.1), three of which have been investigated: the so-called **Damon-Eshbach** (DE) geometry with the magnetization and the wave vector perpendicular to each other, the **Magneto Static Backward Volume** (MSBV) geometry with the two vectors parallel to each other, and an intermediate geometry with the two vectors aligned at an angle of 45°.

### 5.4.1   Continuous Ferromagnetic Film

The standard sample for the observation of spin wave propagation is based on a 20 nm thick ferromagnetic $Ni_{80}Fe_{20}$ thin film with dimensions of 200x200 $\mu$m$^2$ on a GaAs substrate. A 200 nm thick gold CPS with a 10 nm Ti adhesion layer underneath is placed over one edge of the sample, carefully avoiding a shortcut between the signal line and the ground line via the underlying magnetic film. Electron lithography (see appendix C.1), thin film evaporation (see appendix C.3) and subsequent lift-off (see appendix C.4) are used to structure the $Ni_{80}Fe_{20}$ film, while optical lithography (see appendix C.2), thin film evaporation and lift-off are used to create the CPS. An optical micrograph and a cross-section of the sample are shown in fig. 5.3.

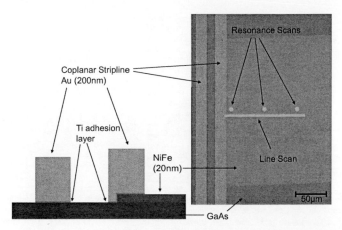

Fig. 5.3: Schematic cross-section (left) and optical micrograph (right) of the continuous film $Ni_{80}Fe_{20}$ sample. On a GaAs substrate, a CPS is patterned over one edge of a 20 nm thick $Ni_{80}Fe_{20}$ film with lateral dimensions of 200x200 $\mu m^2$ while avoiding shortcuts between the signal line and the ground line via the $Ni_{80}Fe_{20}$ film. Two different scanning techniques (see section 6.1) are depicted in the optical micrograph.

## 5.4.2 Structured Interference Samples

In addition to the undisturbed propagation of spin waves, the interference of spin waves was studied. The samples for these experiments have been patterned in a way to create spin waves at different positions and bring these spin waves to interference.

### 5.4.2.1 Leg Structures

This sample is similar to the semi continuous film described in section 5.4.1, with the difference that two legs with a width of 10 $\mu m$ and a distance of 10 $\mu m$ are added on the side that is covered by the CPS. The CPS is now placed over the edges of the legs. Having determined the typical wavelength of spin waves in continuous $Ni_{80}Fe_{20}$ films before, the dimensions have been chosen in the range of these spin wave lengths to allow for interference effects. The signal propagation of the microwave excitation inside the CPS is fast enough to ensure an excitation of spin waves in the two legs with an equal phase. This setup resembles a classic double slit experiment, although some changes like different wave numbers for different directions of propagation from the slits have to be taken into account. An optical micrograph of the structure is shown in fig. 5.4.

39

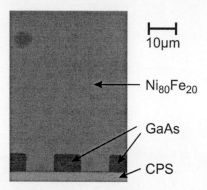

Fig. 5.4: Layout of the sample with legs used for the detection of spin wave interference. The CPS (bottom) and the two legs where spin waves are created can be seen. The dotted grid on the GaAs substrate is caused by the laser hitting the material. Some inevitable resist remnants are burnt in the heat created by the focused laser beam and are visible as points. Each point corresponds to a scanning position in an image scan (see section 6.1.1). The two more pronounced points between the legs mark the position where the beam was parked when no measurements were taken. The $Ni_{80}Fe_{20}$ film was not affected by the laser heat due to its higher thermal conductivity and the lack of resist on its surface. The dark spot in the top left corner and the other sprinkled stains on the $Ni_{80}Fe_{20}$ film, finally, are inevitable but negligible defects from sample production.

### 5.4.2.2 Ring Interferometers

Another possibility to detect spin wave interference is given by ring interferometers. Different to the geometry described before, there is only one source of spin waves on the sample. But then the structure splits up into two branches that reunite again some micrometers away. The spin waves in turn are expected to follow this film structure and interfere with each other where the two branches join again. The idea behind this is two manipulate the the phase of the spin waves on their way through one of the branches of the ring and thus to control the obtained interference. In these structures, this is attempted by adding a topological potential into one branch by patterning a little notch into one branch. It is also imaginable to pin a domain wall here that would affect the phase of the spin waves in this branch [Bay2005-1]. One important difference from the previous two structures is that due to a different patterning procedure, the CPS is not directly placed on top of the $Ni_{80}Fe_{20}$ structures, but with a gap of around 1 $\mu$m besides it. Thus the excitation spectrum of the CPS is different from the previous samples, as the $h_x$ component of the driving field which is largest underneath the signal line cannot be used. This leads to a notably different excitation spectrum (see fig. 4.6) compared to the samples where the film is placed partly underneath the CPS signal line. An optical micrograph of these structures is shown in fig. 5.5.

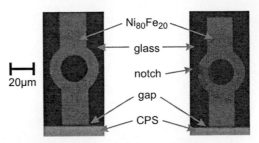

Fig. 5.5: Ring interferometer samples. The small gap between the CPS and the magnetic film leads to a notably different excitation spectrum compared to continuous film samples (see fig. 4.6). In the right structure, a little notch is patterned into one branch of the ring. Due to the use of dry etching for the patterning of this structure, there are some small resist remnants on top of the structures. Using a too intense probe beam, it is thus possible to burn these resist remnants. As an effect, the scanning grid of an image scan can be seen in the right image. The beam was also placed at a position close to the notch for a longer time. Consequently, the sample was destroyed (the foggy region over the notch) and no further measurements could be taken. For the measurements on the left structure, however, the intensity has been chosen smaller, and thus no negative effects to the sample can be found.

The $Ni_{80}Fe_{20}$ structures for this geometry are created by first evaporating a continuous thin $Ni_{80}Fe_{20}$ film on the complete sample, then patterning the structures with optical lithography and argon ion beam etching. Then the CPS is structured using optical lithography, thin film evaporation and subsequent lift-off.

# Chapter 6

# Experimental Results

Using Time Resolved Scanning Kerr Microscopy (TRSKEM), the propagation and interference of spin waves was observed in 20 nm thick $Ni_{80}Fe_{20}$ thin films. The different scanning techniques used here are described in section 6.1.1.

First, the undisturbed propagation of spin waves (see section 6.2) was investigated in samples of a continuous $Ni_{80}Fe_{20}$ film. Using continuous wave (cw) microwave excitation, the resonance frequencies, phase velocities, dispersion relations and damping times were determined for three different geometries of an externally applied in-plane magnetic bias field: the Damon Eshbach (DE) geometry with the magnetization perpendicular to the wave vector, the Magneto Static Backward Volume (MSBV) geometry with the two vectors aligned parallel, and an intermediate 45° geometry with the two vectors forming an angle of 45°. Further on, the group velocity was determined using MicroWave Pulse (MWP) excitation.

Apart from this undisturbed propagation, the interference of spin waves (see section 6.3) was observed as well, in two different samples. First, a structure inspired by a classical double slit experiment was investigated, and the typical interference pattern of a double slit experiment was observed in cw and MWP excitation. In addition, ring interferometer structures were investigated, aiming at a future active control of the interference pattern.

## 6.1   Scanning Techniques

### 6.1.1   Image Scan

As stated in section 4.2, the continuous thin film investigated in this work has a translational invariance parallel to the direction of the CPS. Thus it is not necessary for a continuous thin film to take a scan along this direction. However, as soon as lateral structures come into play, as it is the case for the interference

43

pattern samples, this invariance is no longer found. Therefore we need lateral resolution in both directions along the thin film in order to detect the interference patterns which form in structured samples. For image scans, the sample is scanned in both lateral dimensions by the piezo table. The magnetization is recorded on a grid in $x$ and $y$ direction, and the result is displayed as a 2D image. Image Scans are also important to get an overview of the sample and identify the positions where resonance scans or line scans shall be recorded.

### 6.1.2 Resonance Scan

A resonance scan is performed by focusing the laser on a fixed spot on the sample and running a scan in time. By applying a continuous sinusoidal microwave excitation of a fixed frequency and comparing the amplitudes of the resulting magnetic oscillation, the resonance frequencies of the magnetic film can be evaluated. The results can be directly compared to the resonance frequencies calculated from theory.

### 6.1.3 Line Scan

Apart from running a scan in time only, it is also possible to scan the position of the laser spot on the sample by moving the piezo table on which the sample is mounted. It is advisable to run a scan in a direction perpendicular to the CPS, so that the magnetic oscillation at different distances from the microwave excitation is recorded. By applying a continuous wave excitation, this technique can be used to observe the propagation of a spin wave directly and to determine phase velocities and dispersion relations, while group velocities can be measured by applying a MWP excitation. The wave numbers obtained from these measurements can directly be compared with the results obtained from theory.

### 6.1.4 Image Sequence Scan

Image sequence scans are important for the recording of interfering oscillations, as resolution in time and space is needed. Thus a sequence of image scans is taken at different times relative to the excitation. Combining these image scans, magnetization dynamics and interference patterns can be observed with lateral and time resolution.

## 6.2 Spin Wave Propagation

The propagation of spin waves was observed in continuous $Ni_{80}Fe_{20}$ samples (see section 5.4.1) in three different geometries of an externally applied magnetic in-plane bias field, see fig. 4.1. The resonance frequency for all three geometries depends on the strength of the external magnetic bias field. For a better comparability of the three geometries, the external fields are chosen in a way that the main resonance peak measured in the resonance scans (see section 6.2.1) is close to 4.0 GHz.

A continuous wave (cw) microwave excitation is applied to the sample, and resonance scans are taken in order to determine the shape of the resonance peak and the value of the resonance frequency (see section 6.2.1). After that, line scans around this resonance frequency are taken, from which the phase velocities (see section 6.2.2), wavelengths and dispersion relations (see section 6.2.3) can be extracted. Furthermore, these line scans can be used to determine the amount of damping (see section 6.2.5).

Applying MicroWave Pulses (MWP) to the sample as an alternative excitation, it is also possible to measure group velocities (see section 6.2.4).

### 6.2.1 Resonance Frequency

The measurement of the resonance frequency is the most fundamental technique used to gain information about the propagation of spin waves. The resonance frequency can roughly be estimated by the ferromagnetic resonance frequency $\omega_{FM} = \sqrt{\gamma H_0 (H_0 + M_s)}$ of the homogeneous mode defined by $k = 0$, which is measured in conventional FerroMagnetic Resonance (FMR) experiments. In our setup, however, the magnetization is also excited with wave numbers different from zero, and thus the obtained resonance frequency may differ significantly from the above FMR value.

Fig. 6.1 shows the resonance scans close to the CPS for an applied external magnetic bias field of $\approx$180 Oe (18 mT) for the DE, and $\approx$190 Oe (19 mT) for the MSBV and 45° geometries. The theoretical values are plotted in the same graphs in grey for the DE and MSBV geometries. The dispersion for the 45° geometry lies in between the ones for the DE and MSBV geometries [Ari1999, Ari2000, Mil2003], and results for the 45° geometry are thus expected to be a superposition of DE ans MSBV results. Despite same resonance frequency has been observed in all three geometries, the applied fields are different: The fields are higher for the MSBV and 45° geometries than for the DE geometry. This is equivalent to a higher resonance peak in the DE geometry for equal bias fields.

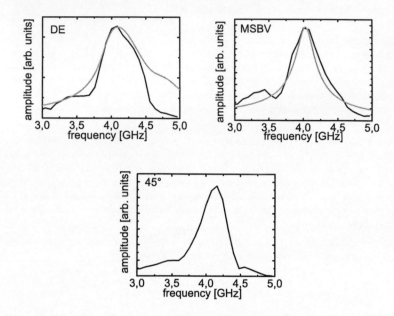

Fig. 6.1: Resonance scans showing (in black) the amplitude of the out-of plane component of the $Ni_{80}Fe_{20}$ film's magnetic oscillation close to the CPS with an external magnetic in-plane bias field of about 180 Oe (18 mT) applied in the DE (left), and 190 Oe (19 mT) in the MSBV (right) and 45° (below) geometries. The bias fields have been chosen so that the resonance frequency is close to 4.0 GHz. The grey curves show fits to the experimental values of the DE and MSBV geometry. Here, the resonance curve in the DE case corresponds to an external field of 177 Oe (17.7 mT), while in the MSBV case it corresponds to a field of 189 Oe (18.9 mT). This behavior is equivalent to an increase of the resonance frequency of the DE geometry compared to the MSBV geometry at the same external field. This is due to the fact that the dynamic susceptibility increases with increasing wave number in the DE case, while it remains nearly constant in the MSBV geometry. This is also the reason for the sharper peak in the MSBV compared to the DE geometry. The susceptibility curve for the 45° geometry is expected to lie between the one for DE and MSBV geometry, but it has actually not been calculated explicitly. However, assuming such a shape of the susceptibility, it can be understood why the frequency of the 45° peak is higher than the one for the MSBV geometry, and sharper than the one for the DE geometry.

The varying observed resonance frequencies can be explained by an interplay of the dynamic susceptibility and the wave number distribution of the driving microwave field, using equations 4.14 and 4.15 for the calculation of the theoretical resonance frequency. It is important to keep in mind that the spin waves in the experiment are excited at a single excitation frequency, but with a distribution of wave numbers (see fig. 4.5), with the wave number spectra being different for the DE and MSBV geometries. The link between driving microwave field and magnetic oscillation is represented by the dynamic susceptibility $\chi(f, k)$. This susceptibility (see fig. 4.4) is different for the DE and MSBV geometries. As the susceptibility depends on the two parameters $f$ and $k$, this means that for each wave number there exists a frequency where the susceptibility has a maximum, i.e. a frequency that is in resonance for this specific wave number. While this line of maximum susceptibility (that is equivalent to the analytic value of the dispersion as calculated in section 4.8) is almost shallow for the MSBV geometry, it rises to higher frequency values with increasing wave number for the DE geometry, with the resonance frequencies for $k = 0$ being identical. This means that, while all wave numbers are resonant to the same frequency in the MSBV geometry, in the DE geometry the higher wave numbers are resonant to higher frequencies.

This leads to two effects that can be observed experimentally and reproduced theoretically. First, the different wave numbers being resonant to different frequencies lead to a broadening of the resonance peak in the DE geometry compared to the MSBV geometry. Second, the resonance peak in the DE geometry is also shifted to higher frequencies due to the steeper susceptibility. As expected, the results for the 45° geometry lie in between those of the DE and MSBV geometries.

It would be possible to record the phase between microwave excitation and magnetic oscillation. However, the meaning of the phase is not really well defined here, as the finite distance between excitation (at the edge of the CPS) and detection (where the laser beam is positioned), together with a finite phase velocity which depends on the applied frequency, make the calculation of the actual phase shift between excitation and response practically impossible.

## 6.2.2 Phase Velocity

After the resonance frequency has been determined, it is possible to go one step further and record the spin wave signature not only at a single position, but along a line perpendicular to the CPS. Recording the magnetic configuration along that line at various times relative to the phase of the microwave driving field, the propagation of the spin waves can be recorded. This propagation is not only measured at the resonance frequency, but for several excitation frequencies below and above the resonance. The phase velocity of the spin waves can be calculated from the slope in the line scans by $v_{Ph} = \frac{\Delta x}{\Delta t}$.

Fig. 6.2 shows line scans taken at different excitation frequencies around the resonance peak in the DE geometry. The line scans for the 45° geometry look very similar and are shown in fig. 6.3. The line scans recorded in the MSBV

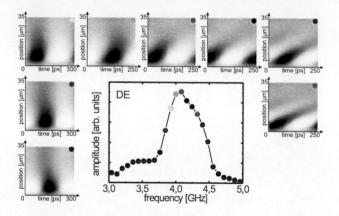

Fig. 6.2: Line scans showing the amplitude of the out-of plane component of the thin film's magnetic oscillation in grey-scale as a function of position and time for an applied external magnetic in-plane bias field of ≈180 Oe (18 mT) in the DE geometry. Colored dots mark the position of the line scan on the resonance scan. The slope in the line scans decreases with increasing frequency, which corresponds to a decreasing phase velocity.

geometry are plotted in the same image and show a very high phase velocity for any applied excitation frequency.

Fig. 6.4 shows the phase velocities calculated from the line scans in figs. 6.2 and 6.3 for the DE and 45° geometries. The phase velocities in the MSBV geometry are very large (close to infinity due to a merely vertical line scan) and are thus not shown in the graph. Phase velocity and wave number are linked to each other by $v_{\mathrm{Ph}} = \frac{2\pi f}{k}$. In the DE and 45° geometries, the decreasing phase velocities correspond to an increase of the wave number $k$ in the dispersion relation, while the merely infinite phase velocity in the MSBV geometry corresponds to a wave number close to zero. Details on the dispersion relation are given in the next section.

## 6.2.3 Dispersion Relation

In section 6.2.2, the phase velocities of spin waves for an applied external field of ≈180-190 Oe (18-19 mT) have been determined for the DE, 45° and MSBV geometries. From the phase velocities, the wave number can easily be calculated by $k = \frac{2\pi f}{v_{\mathrm{Ph}}}$. Plotting the frequency against the wave number, the dispersion relations for the observed spin wave modes are obtained. Fig. 6.5 shows the dispersion relations for the DE and 45° geometries, calculated from the line scans

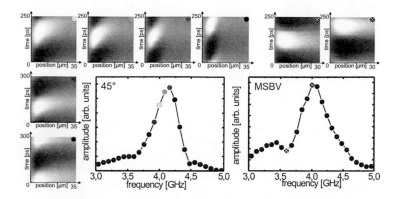

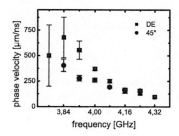

Fig. 6.3: Line scans showing the amplitude of the out-of plane component of the thin film's magnetic oscillation in grey-scale as a function of position and time for an applied external in-plane bias field of ≈190 Oe (19 mT) in the 45° (left) and MSBV (right) geometries. Colored dots mark the position of the line scan on the resonance scan. Similar to the DE geometry, the phase velocity decreases with increasing frequency for the 45° geometry, while the phase velocity is always very large (corresponding to the nearly vertical line scans) for the MSBV geometry, independent from the excitation frequency. Only two line scans are shown for the MSBV geometry, as all the line scans look essentially the same here and do not notably change for any excitation frequency.

Fig. 6.4: Phase velocities $v_{ph}$ for the DE and 45° geometries, extracted from the line scan data shown in figs. 6.2 and 6.3, respectively. The decrease of the phase velocity corresponds to an increase of the wave number.

shown in figs. 6.2 and 6.3, as well as dispersion relations for other values of the applied magnetic bias field in the DE geometry.

In the DE geometry, for a fixed value of the bias field, the observed wave number increases with the excitation frequency (see squares and triangles in fig. 6.5), as

49

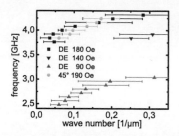

Fig. 6.5: Dispersion of the observed spin waves in the DE and 45° geometry for different values of the bias field. The shape of the dispersion relation in the DE geometry does not notably change for different values of the bias field (squares, triangles). The bias field merely shifts the dispersion curve to different resonance frequencies. For both DE and 45° geometries, the resonance frequency increases with increasing wave number.

expected from analytic calculations (see fig. 4.8). The shape of the dispersion does not change notably, the different bias field mainly shifts the position of the dispersion curve. Keeping the excitation frequency constant and decreasing the bias field, the wave number can thus easily be increased (see black and dark grey data for 180 and 140 Oe (18 and 14 mT), and fig. 6.16). For the 45° geometry, the shape of the dispersion curve is very similar to the DE geometry (see the light grey data for the 45° and the black data for the DE geometry). For a given wave number, the resonance frequency is smaller in the 45° geometry than in the DE geometry (for the presented data, the bias field is slightly higher for the 45° geometry than for the DE geometry, i.e. for the same bias field, the DE dispersion curve would be slightly shifted to even higher frequencies). This can be explained by the different slopes in the susceptibilities for the DE and 45° geometries, as stated in section 6.2.1. Although the susceptibility has not been calculated explicitly for the 45° geometry, it can be stated that the dispersion curve must be located between the ones for the DE and MSBV geometries. The dispersion for the MSBV geometry has not been plotted, as it is mostly centered around zero (see experimental data in fig. 6.6, right).

In sum there are two parameters that can be used to choose a resonant wave number for a given excitation frequency, namely the strength and the direction of the applied in-plane bias field. However, this is only possible if the respective wave number is contained in the wave number spectrum of the driving microwave field.

For the DE geometry, however, the susceptibility has been calculated, and thus it is possible to extract from the theoretical values the dispersion relation expected to be observed in experiment. The respective curves are shown in fig. 6.6 (left).

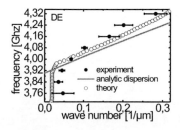 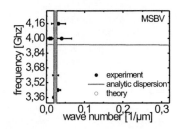

Fig. 6.6: Dispersion of the observed spin waves, for a bias field of ≈ 180 Oe (18 mT) applied in the DE (left) and ≈ 190 Oe (19 mT) applied in the MSBV (right) geometry, obtained analytically (solid line), experimentally (squares) and from the theories described in 4.7 (open circles). The latter take into account the wave number spectrum of the microwave driving field which is neglected in the analytical approach. In the DE geometry, all three dispersion relations, whether obtained from experiment, from theories or analytically, agree quite well with each other. For the MSBV geometry, however, resonances are only found for very small wave numbers in experiment and theories, while a very shallow dispersion results from analytic calculations. The reason for these deviations is found in the very shallow susceptibility and the wave number spectrum of the microwave driving field $h_z$ which mainly contains small wave numbers. These two reasons make it impossible to create spin waves with high $k$ excitation in the MSBV geometry. In the DE geometry, however, a sufficient slope in the susceptibility leads to the creation of isolated excitations with high wave number (see fig. 4.4).

The data obtained experimentally (black) show the behavior already described before. The solid grey line represents the analytic value of the dispersion relation, obtained by setting the denominator of the susceptibility to zero and solving the obtained equation for $k$ and $f$ (see section 4.8). Generally, the experimental values agree quite well with the analytic values, although some deviations can be found that can be explained by taking into account the wave number spectrum of the microwave driving field. According to this analytic dispersion relation, no spin wave frequencies below a threshold frequency are possible. However, in the experiments it has been found that it is possible to excite spin waves with an arbitrary frequency. This excitation of off-resonant spin waves can be modeled theoretically as well, using the method described in section 4.7. In doing so, the susceptibility and the wave number spectrum of the driving microwave field have to be taken into account. The obtained result is plotted into the graph as circles. No analytic formula for this curve can be given, as the numerically obtained wave number spectrum of the microwave driving field is taken into account. The experimental results agree very well with these theoretical expectations.

For the MSBV geometry, however, the differences between analytic dispersion and the results from experiment as well as the theoretically expected values are striking. While a very flat dispersion curve (with slightly decreasing frequency

51

for increasing wave number, in contrast to the DE geometry) is calculated analytically, it is actually not possible to find any dispersion in experiment as well as in the theoretical model. The reason for this is that the susceptibility is very shallow in the MSBV geometry. This means that for a given frequency, the susceptibility will never have an isolated peak for a wave number different from zero. As the wave number spectrum of the microwave driving field is also largest close to zero, this always leads to a maximum of the excitation with a wave number close to zero, no matter what the excitation frequency might be.

In the DE geometry, however, the susceptibility has a large slope, and thus the susceptibility for a constant frequency has a peak isolated from $k = 0$ for all frequencies larger than the analytic resonance frequency for $k = 0$. Thus it is easily possible to excite spin waves with nonzero wave numbers.

### 6.2.4   Group Velocity

The group velocity is defined as the velocity of the propagation of a wave packet (a "group" of waves). The group velocity can be calculated from the dispersion by $v_G = \frac{\partial \omega}{\partial k}$ and is thus given by the slope of the susceptibility. This means that for otherwise constant parameters, the group velocity should be largest for the DE geometry, smaller for the 45° geometry (as its susceptibility, although not calculated, is expected to be flatter than in the DE geometry) and close to zero for the MSBV geometry with its very shallow susceptibility. In fact, it should even be negative, as the susceptibility has a negative slope[1] for small wave numbers $k$.

The group velocity can be determined experimentally by applying not a cw microwave excitation, but short MicroWave Pulses (MWP, see 5.2.2). Fig. 6.7 shows the out-of plane component of the sample's magnetization at different distances from the CPS. The signal has been recorded for an external magnetic bias field of 180 Oe (18 mT) in the DE geometry. From the oscillation at different distances from the CPS the propagation of the spin wave packet can clearly be observed.

The spin wave propagates with a finite velocity. This group velocity can be determined in more detail by another line scan, this time for a period of some ns. The envelope of the wave packet can be obtained by taking the absolute value of the oscillation and then blurring the signal so that the individual oscillations cannot be distinguished any more. The slope of the rising edge of the wave packet

---

[1]To be more precise, it can only be said that group and phase velocity have an opposite sign. As the group velocity is always directed away from the position of excitation, the phase velocity is expected to be negative. For our system, however, it is very hard to detect any of these effects as the spin waves in the MSBV geometry behave different from the analytic dispersion anyway.

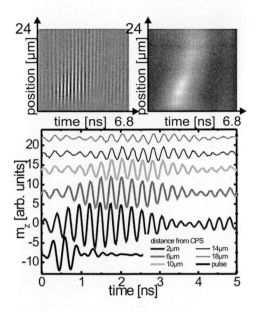

Fig. 6.7: Propagation of a MWP excited spin wave packet, excited at a frequency of 3.60 GHz and an external magnetic bias field of ≈180 Oe (18 mT) in the DE geometry. The excitation frequency has been chosen somewhat below the resonance frequency in order to get higher phase velocities and thus less damping (see fig. 6.13).
Above: Line scan of the oscillation (left) and envelope (right).
Below: The microwave pulse and the oscillations at different distances from the CPS. The propagation of the spin wave (i.e. the onset of the oscillation at different times) can be observed.

can be used to directly measure the group velocity of the wave packet traveling through the thin film.

Figure 6.8 shows the group velocities of spin wave packets in a continuous film for three different geometries of an applied external magnetic bias field. The group velocities are highest for the DE geometry (up to 11 $\mu$m/ns), while they are a little smaller for the 45° geometry, and close to zero for the MSBV geometry. This can be explained with the increased slope of the dispersion relation in the DE geometry, as the group velocity is proportional to the slope of the dispersion, $v_G = \frac{\partial \omega}{\partial k}$. Calculating the group velocity from the analytic dispersion, one would expect a group velocity of around $7 - 7.5$ $\mu$m/ns for the DE geometry and a bias field of 177 Oe (17.7 mT). As the actual dispersion seems to be a little steeper than the analytic solution (see fig. 6.6), the observed group velocities agree quite well with the expected values.

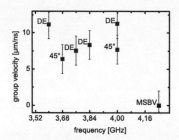

Fig. 6.8: Group velocities for the different geometries and a magnetic bias field of ≈180 Oe (18 mT) in the DE and ≈190 Oe (19 mT) in the MSBV and 45° geometries.

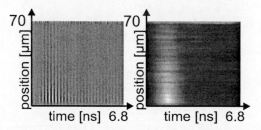

Fig. 6.9: Propagation of a MWP excited spin wave packet, excited at a frequency of 4.24 GHz and an external magnetic bias field of ≈190 Oe (19 mT) in the MSBV geometry, line scan of the oscillation (left) and envelope (right). No real propagation of the wave packet can be found here, the oscillation merely starts and ends with very high phase velocity. Compared to the DE geometry, the spin wave excitation decays notably slower.

For the MSBV geometry, however, it is hard to decide whether the group velocity is zero or infinite. The rising edge of the envelope would correspond to a very high group velocity. However, in the line scans (see fig. 6.9) no real propagation is found, the image rather corresponds to an onset of the oscillation with a very high phase velocity, and a sudden end. This behavior corresponds to a zero group velocity, as expected from the analytic dispersion, although the analytic result should not really be trusted for measurements far away from the analytic dispersion.

## 6.2.5 Damping

There are several ways by which the damping of a spin wave system can be determined. In conventional FMR, the line width is evaluated by $\frac{\Delta\omega}{\omega_0}$. This formula cannot be applied to the frequency peaks measured in this thesis, as a different parameter is measured in this approach (see equation 4.16). However, the damping time can be evaluated using three different approaches.

First, the damping time can be calculated from the theory by the formulas given in equations 4.20 and 4.21. Second, measuring the decay of the oscillation in time at a fixed position upon MWP excitation in time is a method to directly measure the damping time $\tau$. Results are presented in section 6.2.5.1. Third, it is possible to evaluate the attenuation length $\lambda_{att}$, i.e. the decay in a direction perpendicular to the CPS of the amplitude of the cw oscillation (see section 6.2.5.2).

### 6.2.5.1  Damping Time $\tau$

From the resonance frequency calculated analytically in section 4.9, it is possible to derive the damping time $\tau$ by $\tau = \frac{1}{2\pi \mathrm{Im}(f_{res})}$ to $\approx 1.3$ ns. This result does not notably depend on the geometry or the value of the applied bias field.

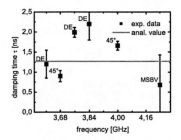

Fig. 6.10: Damping times $\tau$ from the decay of a MWP driven oscillation at a fixed position. The grey line indicates the damping times obtained from the analytic resonance frequency.

The damping time $\tau$ can be directly measured by fitting a damped sine function to the decay of a MWP driven oscillation at fixed position. The results for the wave packets used for the evaluation of the group velocities in fig. 6.8 are shown in fig. 6.10 and compared to the theoretical results. The results generally agree quite well, although some deviations can be found: Smaller damping times can be explained with the increased damping in the film. Larger damping times, however, would correspond to a decreased damping. This might not be the case here; it is more probable that this result is due to the convolution of the decay with the microwave excitation pulse which has a width of $\approx 1$ ns.

### 6.2.5.2 Attenuation Length $\lambda_{att}$

From the line scans used for the calculation of the dispersion relations, it is possible to extract the attenuation length $\lambda_{att}$, i.e. the distance from the driving CPS after which the amplitude of the oscillation is damped to $1/e$, by fitting a damped sine to the oscillation. The attenuation lengths for the investigated DE and 45° geometries are shown in fig. 6.11. As damping is connected to the transport of energy which takes place at the group velocity (which is about 7 $\mu$m/ns), one can try to calculate the damping time $\tau$ by

$$\tau_{v_G} = \frac{\lambda_{att}}{v_G} \, . \tag{6.1}$$

The results are of the same order of magnitude as those from section 6.2.5, although the damping times calculated here differ for different excitation frequencies, in contrast to theoretical expectations (see section 6.2.5.1). Thus the attenuation length might be influenced by a different parameter.

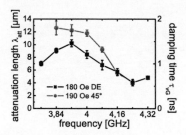

Fig. 6.11: Attenuation lengths $\lambda_{att}$ and damping times $\tau_{v_G}$ for the 180 Oe (18 mT) DE (black) and 190 Oe (19 mT) 45° (grey) geometries. Left axis: attenuation length $\lambda_{att}$, Right axis: damping time $\tau_{v_G}$, calculated for a group velocity of 7$\mu$m/ns.

The attenuation lengths (see fig. 6.11) decrease with increasing frequency. The same behavior has also been found for the phase velocities in section 6.2.2. A closer look at the data reveals that the two variables are proportional to each other. The proportionality constant has the dimension of a time and shall be called the attenuation time $\tau_{att}$ here, defined by

$$\tau_{att} = \frac{\lambda_{att}}{v_{Ph}} \, . \tag{6.2}$$

The phase velocities, attenuation lengths and attenuation times are plotted in fig. 6.12. The attenuation times $\tau_{att}$ are at least one order of magnitude smaller

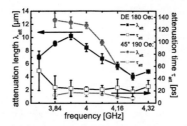

Fig. 6.12: Attenuation lengths $\lambda_{att}$ (left axis) and attenuation times $\tau_{att}$ (right axis) for the 180 Oe (18 mT) DE (black) and 190 Oe (19 mT) 45° (grey) geometries. The attenuation times are almost constant within one geometry, although with rather large errors. The size of the error bars is so large as the attenuation time $\tau_{att}$ depends (via the phase velocity) on the wave number $k$ that contains rather large errors itself.

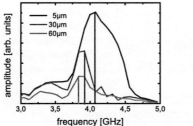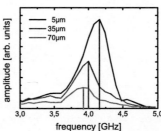

Fig. 6.13: Resonance scans showing the amplitude of the out-of plane component of the thin film's magnetic oscillation at different distances from the CPS with an external magnetic in-plane bias field of about 180 Oe (18 mT) applied in the DE (left) and 190 Oe (19 mT) in the 45° (right) geometry. Graphs for different distances are not to scale. The main peak decreases in frequency with increasing distance from the CPS. The shift in the observed peak does not mean a real shift in the resonance frequency, this would hardly be possible. This is rather due to the different attenuation lengths for the different frequencies.

than the damping time $\tau$ (which can be calculated to ≈1.3 ns), thus they are not directly connected to each other.

It is not easy to understand why the attenuation length should be proportional to the phase velocity. Assuming the dissipation of energy to be responsible for the attenuation of the oscillation, one would expect the group velocity to govern the attenuation length. Calculating the damping time from the attenuation length and the group velocity, the result is of the expected magnitude, but not constant for different excitation frequencies, in contrast to theoretical expectations. One should also keep in mind that the attenuation length has been determined from a cw excited oscillation. This means that a group velocity is not really present. The phase velocity, in contrast, is not linked to the transport of energy. Nevertheless, they $v_{Ph}$ and $\lambda_{att}$ are found to be proportional to each other, a fact that indicates a physical relation of the two.

The concept of the attenuation time can also be used to explain another effect: The resonance scans taken for the determination of the resonance frequency have always been recorded very close to the CPS. However, if the same resonance scans are recorded further away from the CPS, the position of the resonance peak will be shifted to lower frequencies (see fig. 6.13). If the attenuation time $\tau_{att}$ is regarded as a physical variable that governs the damping of a cw excited wave according to equation 6.2, the different attenuation lengths $\lambda_{att}$ can easily be explained by the different phase velocities for the different frequencies.

## 6.3   Spin Wave Interference

Having investigated the undisturbed propagation of spin waves, it is now possible to go one step further, and examine the interaction of spin waves with each other and to try to manipulate the resulting interference patterns. The understanding and control of the formation of the interference is a necessary prerequisite for the construction of logical devices based upon the interaction and propagation of spin waves.

In a first approach, the interference of spin waves is investigated in a structure motivated by a classical double slit experiment (see section 6.3.1). In this very fundamental setup, two spin waves with equal phase are brought to interfere, and the resulting interference pattern is observed.

The next step would be to actively control the interference pattern, possibly by controlling the phase of the two spin waves. A feasible setup for an experiment like this are ring interferometers (see section 6.3.2), where the phase of one spin wave could be controlled by locally applying an external electric potential [Sch2006] or pinning a domain wall in one branch of the ring [Her2004].

### 6.3.1   Leg Structures

In an experiment motivated by classical double slit experiments, the interference patterns of spin waves in a patterned 20 nm thick $Ni_{80}Fe_{20}$ film (see section 5.4.2.1) are investigated. Spin waves are created close to the CPS, inside the two "legs" of the structure, and propagate towards an area of continuous film where they are brought to interference. The propagation of the microwave excitation inside the CPS is fast enough to cause an in-phase excitation of the spin waves

58

in the two legs. Results of a typical interference experiment for continuous wave (cw) excitation in the DE geometry are presented in fig. 6.14. The frequency has been chosen somewhat below the resonance frequency in order to achieve a longer propagation of the spin wave packet (see fig. 6.13).

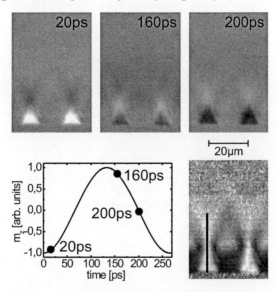

Fig. 6.14: cw spin wave interference in the DE geometry. Top: Snapshots of the out-of plane component of the magnetization at different phases of the microwave excitation (bottom left). Bottom right: Amplitude distribution of the interference pattern. The black line marks the position of the line scans in fig. 6.20. The magnetic bias field is 170 Oe (17 mT), and the excitation frequency 3.76 GHz. For the layout of the sample, see fig. 5.4.

The spin waves first propagate inside the legs towards the continuous film region. From the position where the two legs are connected to the continuous films, the spin waves radiate into the film. Note that there are deviations from a circular propagation (which would be expected from Huygens principle) due to the different dispersion and different wave numbers $k$ for the different directions of propagation of the diffracted spin waves. Compared to the undisturbed propagation, we do not observe a plane wave front, but several finger-like structures that reach out into the film and form a typical cross-like structure. This pattern becomes more obvious when looking at the local amplitude distribution of the interference pattern (see fig. 6.14, bottom right). The cross structure can be identified as regions with low oscillation amplitude, i.e. the nodal lines of the interference pattern.

It has been shown in fig. 6.5, that it is possible to control the wave number of the spin wave by changing the external magnetic bias field. It should thus be possible to manipulate the wave numbers and as a consequence the interference pattern by changing the external magnetic bias field. The results of this are shown in fig. 6.15, where snapshots of the magnetization dynamics for different bias fields are presented.

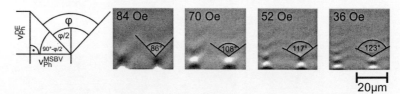

Fig. 6.15: Interference with high $k$ vector, 3.44 GHz cw excitation, various bias fields. The sections shown are smaller than the ones in fig. 6.14 and cover the area where the legs open into the continuous film region. The opening angle of the diffracted spin waves can be used to calculate the wave numbers in the direction perpendicular to the direction of propagation. Due to the dispersion in the DE geometry, the wave number $k$ rises with decreasing bias field for a constant excitation frequency (see fig. 4.8). This can also be understood in a way that the in-plane bias field and the increasing in-plane demagnetizing field due to the increasing wave number always add up to the same effective magnetic field for a constant excitation frequency. The sketch to the left shows the geometry used to calculate the wave numbers in the DE and MSBV directions using equation 6.7.

The wave number $k$ increases with decreasing bias field. This can be understood in two ways, by a consideration of the dispersion relation or by looking at the effective magnetic fields. Assume an excitation frequency $f_0$ and a bias field $H_0$ for which the resonant wave number is 0, i.e. the excitation frequency is identical to the resonance frequency for $k=0$. Now the bias field shall be decreased by an amount $\Delta H$. In the dispersion picture, this means that the dispersion is shifted to lower values, and thus the crossing point of the excitation frequency and the dispersion is shifted to a wave number $k_1$. This means that the wave number $k_1$ is now resonant to the excitation frequency $f_0$, rather than the wave number $k=0$ (see fig. 6.16).

Considering the effective magnetic field, a method common in conventional FMR, the lowering of the external bias field $H_0$ to $H_0 - \Delta H$ means that the lack of external magnetic field has to be made up for by another source of magnetic field, in this case an increase of the in-plane demagnetizing field $H_{d,x}$. In other words, this means that the spin wave with the wave number $k_1$ must create an in-plane demagnetizing field of the amount $\Delta H$ that makes up for the reduced external in-plane bias field. External bias field and demagnetizing field always sum up to the same effective field $H_0$ for a fixed excitation frequency $f_0$.

Looking at the analytic formula for the resonance frequency, this simple assump-

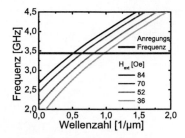

Fig. 6.16: Change of the resonant wave number with the applied bias field. The lowering of the bias field leads to a shift of the dispersion relations to lower frequencies. Keeping the excitation frequency constant, this leads to an increase of the resonant wave number, which is given by the crossing point of the excitation frequency and the dispersion relation.

tion is found to be approximately correct. Going back to the dispersion picture and equation 4.18, neglecting damping and exchange interaction, the following term is obtained for the effective magnetic field underneath the square root (the value of this root dominates the resonance frequency):

$$H_{\text{eff}} = 4H_{\text{d,x}}\left(M_s - H_{\text{d,x}}\right) + 4H_0\left(M_s + H_0\right) , \qquad (6.3)$$

where the relation $H_{\text{d,x}} + H_{\text{d,z}} = M_s$ has been taken into account. Assuming a change of $H_0$ to $H_0 - \Delta H$, this means a change of the above effective field to

$$H_{\text{eff}} = 4H_{\text{d,x}}\left(M_s - H_{\text{d,x}}\right) + 4H_0\left(M_s + H_0\right) - 4\Delta H\left(M_s + 2H_0 - \Delta H\right) \qquad (6.4)$$

If the demagnetizing field $H_{\text{d,x}}$ is now increased by the same amount $\Delta H$, this leads to

$$
\begin{aligned}
H_{\text{eff}} =\ & 4H_{\text{d,x}}\left(M_s - H_{\text{d,x}}\right) + 4H_0\left(M_s + H_0\right) + \\
& +4\Delta H\left(M_s + 2H_{\text{d,x}} + \Delta H\right) - 4\Delta H\left(M_s + 2H_0 - \Delta H\right)
\end{aligned} \qquad (6.5)
$$

As the two resulting frequencies are supposed to be equal, this means that the two corrections have to add up to zero, namely

$$4\Delta H\left(M_s + 2H_{\text{d,x}} + \Delta H\right) - 4\Delta H\left(M_s + 2H_0 - \Delta H\right) \approx 0 \qquad (6.6)$$

Since the saturation magnetization $M_s$ is at least one order of magnitude larger than the other fields appearing here, the substitution of the bias with the demagnetizing field can be regarded as a good approximation, as the above sum then is approximately zero. This is still an approximation and no exact relation. The reason for this is that the external bias field and the in-plane demagnetizing field are perpendicular to each other, which means that they do not add up directly. However, this relation commonly used in conventional FMR is still a good approximation. The values of the respective bias-, demagnetizing and effective magnetic fields for the measurements presented in fig. 6.16 are plotted in fig. 6.17.

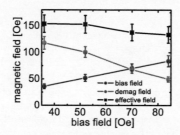

Fig. 6.17: Plot of the bias, demagnetizing and effective magnetic fields for various bias fields and a constant excitation frequency of 3.44 GHz. The demagnetizing field makes up for the changes of the bias field, thus the two always add up to approximately the same effective field.

Turning back to the data presented in fig. 6.16, it can be stated that with decreasing wave length, the distance of the two legs becomes too large to observe a proper interference pattern. However, this also means that there is still much space for miniaturization left, making logical elements based upon spin wave interference even more attractive. Keeping in mind the group velocity of just below 10 $\mu$m/ns, miniaturization is quite important for future applications. For the given group velocity, a distance of 10 $\mu$m corresponds to a response time of at least 1 ns for future logic elements based upon spin waves.

The opening angle of the diffracted spin waves from fig. 6.15 as well as the calculated wave numbers of the spin waves propagating in the DE geometry and the diffracted spin waves in a direction perpendicular to the propagation, are shown in fig. 6.18. As expected for the DE geometry and explained above, the wave number of the propagating spin wave decreases with increasing magnetic bias field. The wave number of the diffracted spin wave can be calculated by

$$k^{\mathrm{MSBV}} = k^{\mathrm{DE}} \cdot \tan\left(90° - \phi/2\right) \qquad (6.7)$$

with $\phi$ being the opening angle as plotted in fig. 6.15. In the same figure, the geometry used to derive the above formula is shown. The wave number of the diffracted spin wave stays more or less constant for varying bias fields. This can be understood as the properties of this spin wave resemble the MSBV geometry (wave vector parallel to the magnetization). However, the geometry is not identical to the MSBV geometry, as in this case here there is an additional periodical in-plane demagnetizing field perpendicular to the wave vector which is due to the wave number $k^{DE}$ of the propagating spin wave. As stated above, this in-plane demagnetizing field makes up for the decreased external bias field. Thus the diffracted spin wave "sees" the same effective magnetic field for all values of an externally applied bias field, which results in the same resonant wave number of the diffracted spin wave in the MSBV geometry for all applied fields.

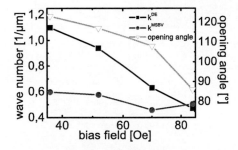

Fig. 6.18: Plot of the opening angles (triangles) and the wave numbers of the propagating spin wave (squares) and the diffracted spin waves (circles) in a direction perpendicular to the propagation.

Finally, it is possible to perform MicroWave Pulse (MWP) driven experiments on this sample. In fig. 6.19, the magnetization configuration of the interference pattern is shown at different times relative to the onset of the microwave excitation pulse. Here, the creation of the spin wave close to the CPS (first 500 ps), its propagation in the legs (500-1200 ps), and the following radiation into the continuous film region and the formation of the interference pattern can be seen. As in the cw excitation case, the formation of a cross shaped structure is found. However, the cross structure does not appear to be located at a fixed position, but travels across the film with the group velocity.

After the MWP is switched off ($\approx$1000 ps), the coherence of the oscillation inside the legs is lost. Instead of a single wave front, spin wave resonances inside the legs are found. These standing spin waves also radiate into the continuous film region.

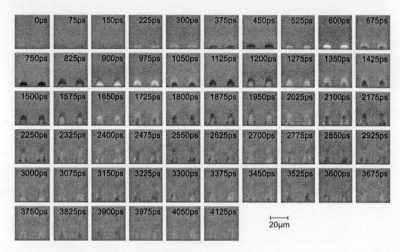

Fig. 6.19: Images of the interference of a MWP driven spin wave packet. The images have been taken at a frequency of 2.88 GHz and a bias field of ≈100 Oe (10 mT) in the DE geometry. The pulse shape and a line scan through the left leg are shown in fig. 6.20.

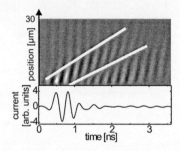

Fig. 6.20: Top: Line scans through one leg of the interference sample, taken at the position marked in fig. 6.14, for MWP driven Interference, as shown in fig. 6.19. The first wave packet (between white lines) is excited by the microwave pulse. Broadening of the spin wave packet due to dispersion can be observed. The second wave packet is due to the decay of resonant spin wave excitations inside the legs. Bottom: MWP used for the excitation of the spin wave packet.

This behavior can be further investigated by recording line scans through one leg, as shown in fig. 6.20. Here, line scans have been recorded as a function of time. Two spin wave packets can be seen propagating into the continuous film region, a first one driven by the MWP excitation, where notable broadening of the spin wave packet due to dispersion can also be found. The group velocity of this spin wave packet can be determined to $6.3 \pm 0.5$ $\mu$m/ns, in good agreement with the theoretical expectations (see section 6.2.4). After this first spin wave packet, a second one occurs. A comparison to the images shown in fig. 6.19 reveals that these additional spin waves are due to the decay of standing spin waves inside the sample's legs: As the internal fields inside the legs differ from those of a continuous film due to the demagnetizing fields of the edges, standing spin waves similar (but not identical) to those in rectangular structures can be excited. These spin waves are still present inside the legs when the MWP is switched off, and the decaying oscillation radiate spin waves into the continuous film.

## 6.3.2   Ring Interferometers

For another type of interference experiments, ring interferometers (see section 5.4.2.2) are used which differ from the leg structures treated above. While in the leg structures spin waves are created at two different positions, spin waves are created at one position only in a ring interferometer. These spin waves are divided up into two branches and are then brought back to interfere. The experiments with the ring interferometers do not aim at the observation of the interference pattern. Moreover, these ring interferometers are intended for the control of the spin wave interference (i.e. amplification or annihilation) by external fields. The experiments presented in this section are intended as a first "proof of principle".

As mentioned in section 5.4.2.2, the ring interferometers were not placed partly under the CPS, but with a narrow gap besides the CPS. Thus it was not possible to observe an oscillation in the DE geometry which relies on the $h_x$ microwave driving field of the CPS which is mainly located underneath the signal line. It was only possible to observe spin waves in the MSBV geometry with its very low wave numbers.

In fig. 6.21, the out-of plane magnetization of a ring interferometer in the MSBV geometry and a magnetic bias field of $\approx$200 Oe (20 mT) is shown for an excitation frequency of 4.64 GHz. It is clearly visible that the spin wave indeed follows the shape of the ring interferometer, as can be seen in the amplitude distribution of the oscillation (see fig. 6.22).

In a first attempt to manipulate the interference of these ring interferometers, a notch was structured in one of the branches of the ring (see fig. 5.5). However, as only MSBV geometry could be used with the very large wave lengths, the effect of the notch was too small to create a significant effect: the wave lengths typically need to be in the range of the manipulating potential in order to cause a sizable effect.

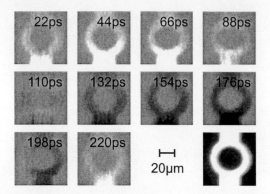

Fig. 6.21: Rings in the MSBV geometry, driven by a cw microwave excitation of 4.64 GHz at a magnetic bias field of ≈200 Oe (20 mT) in the MSBV geometry. The lower right square shows the topography signal of the structure. An optical micrograph of the structures is shown in fig. 5.5.

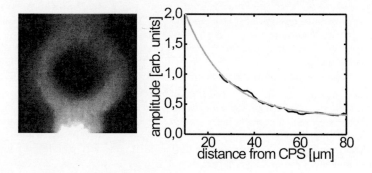

Fig. 6.22: Amplitude distribution inside the MSBV excited ring interferometer shown in fig. 5.5 (left), and this amplitude as a function of distance to the CPS (right). The spin wave can still be observed behind the ring structure, making these ring interferometers suited for further interference experiments.

In the future, these ring interferometers could be used to modify the dispersion of a spin wave inside one branch by locally applying external electric [Sch2006] or magnetic fields and thus modifying the resonance output.

In these first experiments, it could be shown that the spin waves follow the structure of the ring interferometer and still be observed at the region behind the ring where the interference takes place. Using slightly modified interferometers that are at least partly placed underneath the CPS, it will be possible to observe spin wave interference in the DE geometry with smaller wave lengths in these samples. For these smaller wave lengths, it should also be possible to manipulate the phase of the spin wave and thus control the interference.

# Chapter 7

# Summary & Outlook

In this thesis, the propagation of spin waves in thin ferromagnetic $Ni_{80}Fe_{20}$ films has been observed using TRSKEM for different geometries of an externally applied magnetic bias field. Resonance frequencies, phase velocities, dispersion relations, group velocities and damping times have directly been measured and were in good agreement with a theoretical model based upon linearization of the LLG equation. Further on, the interference of spin waves was observed in a layout motivated by a classical double slit experiment. The next step can now be to construct logical elements that rely upon the interference of spin waves.

In a first step towards this goal, ring interferometer samples have been investigated. The idea here is to split a spin wave up into two, and rejoin them again later. While they are split up, the phase of one spin can be manipulated, so that the interference of the two rejoining spin waves switches from amplification to annihilation. There are several ideas of how the phase of the spin waves could be controlled, e.g. by sending the spin wave through a Bloch wall [Bay2005-1, Her2004] or by applying local magnetic [Bay2006, Dem2004] or electric [Sch2006, Wei2007] fields. For special materials, even changes of the magnetic structure are possible [Lot2004] due to applied electric fields. Only recently, there have been reports on controlling spins by using light [Kim2005, Sta2007-1, Sta2007-2]. Combining this new technique with spin wave interference or focusing [Vee2006], it might be possible to create ultrafast logic elements.

Given these new possibilities and the already realized applications of spins like MRAM, there will certainly be many interesting developments in spintronics and magnonics in the next years.

# Appendix A

# Theoretical Supplement

## A.1 Fourier Representation of Magnetic Fields

An arbitrary magnetization distribution $\mathbf{M}(\mathbf{r})$ can be written in terms of a plane wave expansion

$$\mathbf{M}(\mathbf{r}) = \sum_k \mathbf{m_k}(z)\, e^{i\mathbf{k}\cdot\mathbf{r}} . \tag{A.1}$$

In the case of undamped propagating spin waves, we only have one Fourier component with one wave vector.

Furthermore, also the magnetic microwave driving fields created by the CPS can be treated in a Fourier representation method. Using standard em-packages (e.g. Sonnet or Empire), it is possible to calculate the spatial distribution of the magnetic field created by a CPS. Using Fourier transformation (e.g. using MatLab), it is possible to calculate from this data the wave number distribution of the microwave driving fields. The Fourier expansion of these fields can be interpreted in a way analogous to equation A.1.

## A.2 Purely 1D Calculation of Spin Wave Demagnetizing Fields

In this section, a very simple 1D approach of calculating the demagnetizing fields of a spin wave is presented. However, these results are not used for further calculations. They are merely intended as a simple attempt to explain the formulas for the demagnetizing field given in [Har1968].

Starting from the magnetization in the DE case $\mathbf{M}^{\mathrm{DE}} = \begin{pmatrix} m_x e^{i(kx-\omega t)} \\ M_s \\ m_z e^{i\left(kx-\frac{\pi}{2}-\omega t\right)} \end{pmatrix}$ we

calculate the magnetic volume charges by $\lambda_v = -\nabla \cdot \mathbf{M} = -ikm_x e^{i(kx-\omega t)}$. From these volume charges, the demagnetization field can be calculated by[1]

$\mathbf{H}_{\mathrm{d,x}} = \int \lambda_v = \begin{pmatrix} m_x e^{i(kx-\omega t)} \\ 0 \\ 0 \end{pmatrix}$. This means that the demagnetizing field is

proportional to the magnetization and does not depend on the wave number. The magnetic surface charges $\sigma_s$, in turn, can be calculated by

$$\sigma_s = \mathbf{M} \cdot \mathbf{n} = \begin{pmatrix} 0 \\ 0 \\ m_z e^{i\left(kx-\frac{\pi}{2}-\omega t\right)} \end{pmatrix}$$

and, analogously to a plate capacitor, lead to a demagnetizing field

$$\mathbf{H}_{\mathrm{d,z}} = \sigma_s/d = \begin{pmatrix} 0 \\ 0 \\ \frac{m_z}{d} e^{i\left(kx-\frac{\pi}{2}-\omega t\right)} \end{pmatrix}.$$

The calculations in the MSBV case are the same, except for the fact that there exists no in-plane demagnetizing field, as the magnetization in the $x$ direction is constant within our approximations.

Note that all the calculated demagnetizing fields do not depend on the wave number $k$, which cannot be true. The reason for this over simplified result is that the stray fields outside the sample are neglected in the calculations. However, these results can be regarded as a good approximation in the limits of small (MSBV case) or large (DE case) wavenumbers, where stray fields outside the sample are very small. In fact, one finds that the limits of the formula in [Har1968] exactly agree with the results of this very simple and pictorial calculation.

## A.3 Calculation of the Dynamic Susceptibility

The dynamic susceptibilities $\chi$ are calculated for the DE and MSBV geometries by linearizing the LLG equation, an approach similarly followed by [Hil1990, McM1998, Wol2004]. An advantage of this approach is that it is rather pictorial. A different approach, being more theoretical, can be found in [Kal1986], for example.

---

[1]This is essentially the same as calculating the field of a charged infinite plane in three dimensions. Thus there is no factor that induces a dependence on the distance from the charge, as it would be a dependence $\frac{1}{r^2}$ for a point charge in three dimensions.

The following sections present the calculations only, for more information on the analytic approach followed here see chapter 4.

## A.3.1   DE Geometry

In the DE geometry, the external magnetic field is

$$
\mathbf{H}_{\text{ext}}^{\text{DE}} = \begin{pmatrix} h_x e^{-i\omega t} \\ H_0 \\ h_z e^{-i\omega t} \end{pmatrix} \tag{A.2}
$$

with the static external magnetic bias field $\mathbf{H}_0$ and the high frequency driving fields $h_x$ and $h_z$.

In the following, we assume a continuous magnetic film with a magnetization distribution that can be described by a single plane wave ('spin wave') with a single wave number $k$. Damping is thus neglected for the calculation of magnetic fields. It only comes into play with the damping parameter $\alpha$ in the LLG-equation, where it causes a broadening of the resonance peak and a small downshift of the resonance frequency.

Neglecting the small deviations from the saturation magnetization in the $y$-direction, the magnetization vector can thus be written as

$$
\mathbf{M}^{\text{DE}} = \begin{pmatrix} m_x e^{i(kx-\omega t)} \\ M_s \\ m_z e^{i\left(kx-\frac{\pi}{2}-\omega t\right)} \end{pmatrix} . \tag{A.3}
$$

The exchange field is

$$
\mathbf{H}_{\text{exch}}^{\text{DE}} = \frac{2A}{\mu_0 M_s^2} \nabla^2 \mathbf{M} = \frac{2A}{\mu_0 M_s^2} \begin{pmatrix} \frac{\partial^2 m_x e^{i(kx-\omega t)}}{\partial x^2} \\ 0 \\ \frac{\partial^2 m_z e^{i\left(kx-\frac{\pi}{2}-\omega t\right)}}{\partial x^2} \end{pmatrix} \tag{A.4}
$$

According to equation 4.1 the demagnetizing field in the DE case can be written as

$$
\mathbf{H}_{\text{d}}^{\text{DE}} = - \begin{pmatrix} m_x e^{i(kx-\omega t)} \frac{kd-1+e^{-kd}}{kd} \\ 0 \\ m_z e^{i\left(kx-\frac{\pi}{2}-\omega t\right)} \frac{1-e^{-kd}}{kd} \end{pmatrix} . \tag{A.5}
$$

Putting this into equation the LLG equation 3.1, one has

$$
\begin{pmatrix} \frac{dm_x e^{i(kx-\omega t)}}{dt} \\ 0 \\ \frac{dm_z e^{i\left(kx-\frac{\pi}{2}-\omega t\right)}}{dt} \end{pmatrix} = -\mu_0\gamma \begin{pmatrix} m_x e^{i(kx-\omega t)} \\ M_s \\ m_z e^{i\left(kx-\frac{\pi}{2}-\omega t\right)} \end{pmatrix} \times
$$

$$
\times \left[ \left( \begin{pmatrix} h_x e^{-i\omega t} \\ H_0 \\ h_z e^{-i\omega t} \end{pmatrix} + \frac{2A}{\mu_0 M_s^2} \begin{pmatrix} \frac{\partial^2 m_x e^{i(kx-\omega t)}}{\partial x^2} \\ 0 \\ \frac{\partial^2 m_z e^{i\left(kx-\frac{\pi}{2}-\omega t\right)}}{\partial x^2} \end{pmatrix} - \begin{pmatrix} m_x e^{i(kx-\omega t)}\frac{kd-1+e^{-kd}}{kd} \\ 0 \\ m_z e^{i\left(kx-\frac{\pi}{2}-\omega t\right)}\frac{1-e^{-kd}}{kd} \end{pmatrix} \right) \right] +
$$

$$
+ \; \frac{\alpha}{M_s} \begin{pmatrix} m_x e^{i(kx-\omega t)} \\ M_s \\ m_z e^{i\left(kx-\frac{\pi}{2}-\omega t\right)} \end{pmatrix} \times \begin{pmatrix} \frac{dm_x e^{i(kx-\omega t)}}{dt} \\ 0 \\ \frac{dm_z e^{i\left(kx-\frac{\pi}{2}-\omega t\right)}}{dt} \end{pmatrix} \tag{A.6}
$$

which reduces to

$$
-i\omega \begin{pmatrix} m_x e^{i(kx-\omega t)} \\ 0 \\ m_z e^{i\left(kx-\frac{\pi}{2}-\omega t\right)} \end{pmatrix} = -\mu_0\gamma \begin{pmatrix} m_x e^{i(kx-\omega t)} \\ M_s \\ m_z e^{i\left(kx-\frac{\pi}{2}-\omega t\right)} \end{pmatrix} \times \tag{A.7}
$$

$$
\times \begin{pmatrix} h_x e^{-i\omega t} - \left(\frac{2Ak^2}{\mu_0 M_s^2} + \frac{kd-1+e^{-kd}}{kd}\right) m_x e^{i(kx-\omega t)} \\ H_0 \\ h_z e^{-i\omega t} - \left(\frac{2Ak^2}{\mu_0 M_s^2} + \frac{1-e^{-kd}}{kd}\right) m_z e^{i\left(kx-\frac{\pi}{2}-\omega t\right)} \end{pmatrix}
$$

$$
- \; \frac{i\omega\alpha}{M_s} \begin{pmatrix} m_x e^{i(kx-\omega t)} \\ M_s \\ m_z e^{i\left(kx-\frac{\pi}{2}-\omega t\right)} \end{pmatrix} \times \begin{pmatrix} m_x e^{i(kx-\omega t)} \\ 0 \\ m_z e^{i\left(kx-\frac{\pi}{2}-\omega t\right)} \end{pmatrix}
$$

Calculating the cross product, keeping only terms linear in $m_x$, $m_z$, $h_x$, $h_z$ and omitting the time dependence $e^{-i\omega t}$, one finally obtains:

$$
-i\omega \begin{pmatrix} m_x e^{ikx} \\ 0 \\ m_z e^{i\left(kx-\frac{\pi}{2}\right)} \end{pmatrix} = -\mu_0\gamma \begin{pmatrix} M_s h_z - \left(\frac{2Ak^2}{\mu_0 M_s} + M_s\frac{1-e^{-kd}}{kd} + H_0 - i\alpha\frac{\omega}{\mu_0\gamma}\right) m_z e^{i\left(kx-\frac{\pi}{2}\right)} \\ 0 \\ -M_s h_x + \left(\frac{2Ak^2}{\mu_0 M_s} + M_s\frac{kd-1+e^{-kd}}{kd} + H_0 - i\alpha\frac{\omega}{\mu_0\gamma}\right) m_x e^{ikx} \end{pmatrix}
$$

This can be written as a system of two coupled linear equations:

$$
m_x e^{ikx} = -i\frac{\mu_0\gamma}{\omega}\left[ M_s h_z - \left(\frac{2Ak^2}{\mu_0 M_s} + M_s\frac{1-e^{-kd}}{kd} + H_0 - i\alpha\frac{\omega}{\mu_0\gamma}\right) m_z e^{i\left(kx-\frac{\pi}{2}\right)} \right]
$$

$$
m_z e^{i\left(kx-\frac{\pi}{2}\right)} = -i\frac{\mu_0\gamma}{\omega}\left[ -M_s h_x + \left(\frac{2Ak^2}{\mu_0 M_s} + M_s\frac{kd-1+e^{-kd}}{kd} + H_0 - i\alpha\frac{\omega}{\mu_0\gamma}\right) m_x e^{ikx} \right]
$$

Putting the first equation into the second, one obtains an equation for $m_z e^{i\left(kx-\frac{\pi}{2}\right)}$:

$$
\begin{aligned}
m_z e^{i\left(kx-\frac{\pi}{2}\right)} =\ & i\frac{\mu_0\gamma}{\omega}M_s h_x \\
& - \frac{(\mu_0\gamma)^2}{\omega^2}M_s h_z \left(\frac{2Ak^2}{\mu_0 M_s} + M_s\frac{kd-1+e^{-kd}}{kd} + H_0 - i\alpha\frac{\omega}{\mu_0\gamma}\right) \\
& + \frac{(\mu_0\gamma)^2}{\omega^2}\left(\frac{2Ak^2}{\mu_0 M_s} + M_s\frac{1-e^{-kd}}{kd} + H_0 - i\alpha\frac{\omega}{\mu_0\gamma}\right)\cdot \\
& \cdot\left(\frac{2Ak^2}{\mu_0 M_s} + M_s\frac{kd-1+e^{-kd}}{kd} + H_0 - i\alpha\frac{\omega}{\mu_0\gamma}\right)m_z e^{i\left(kx-\frac{\pi}{2}\right)}
\end{aligned}
$$

from which one can derive the equation

$$
m_z = \frac{M_s\left(\frac{2Ak^2}{\mu_0 M_s} + M_s\frac{kd-1+e^{-kd}}{kd} + H_0 - i\alpha\frac{\omega}{\mu_0\gamma}\right)h_z - i\frac{\omega}{\mu_0\gamma}M_s h_x}{\left(\frac{2Ak^2}{\mu_0 M_s} + M_s\frac{1-e^{-kd}}{kd} + H_0 - i\alpha\frac{\omega}{\mu_0\gamma}\right)\left(\frac{2Ak^2}{\mu_0 M_s} + M_s\frac{kd-1+e^{-kd}}{kd} + H_0 - i\alpha\frac{\omega}{\mu_0\gamma}\right) - \frac{\omega^2}{(\mu_0\gamma)^2}}e^{-i\left(kx-\frac{\pi}{2}\right)}.
$$

$$(A.8)$$

From this equation, the susceptibilities $\chi_{zx}^{\mathrm{DE}}$ and $\chi_{zz}^{\mathrm{DE}}$ can easily be identified as

$$
\chi_{zx}^{\mathrm{DE}} = \frac{m_z}{h_x} = \tag{A.9}
$$

$$
= \frac{-i\frac{\omega}{\mu_0\gamma}M_s}{\left(\frac{2Ak^2}{\mu_0 M_s} + M_s\frac{1-e^{-kd}}{kd} + H_0 - i\alpha\frac{\omega}{\mu_0\gamma}\right)\left(\frac{2Ak^2}{\mu_0 M_s} + M_s\frac{kd-1+e^{-kd}}{kd} + H_0 - i\alpha\frac{\omega}{\mu_0\gamma}\right) - \frac{\omega^2}{(\mu_0\gamma)^2}}e^{-i\left(kx-\frac{\pi}{2}\right)}
$$

and

$$
\chi_{zz}^{\mathrm{DE}} = \frac{m_z}{h_z} = \tag{A.10}
$$

$$
= \frac{M_s\left(\frac{2Ak^2}{\mu_0 M_s} + M_s\frac{kd-1+e^{-kd}}{kd} + H_0 - i\alpha\frac{\omega}{\mu_0\gamma}\right)}{\left(\frac{2Ak^2}{\mu_0 M_s} + M_s\frac{1-e^{-kd}}{kd} + H_0 - i\alpha\frac{\omega}{\mu_0\gamma}\right)\left(\frac{2Ak^2}{\mu_0 M_s} + M_s\frac{kd-1+e^{-kd}}{kd} + H_0 - i\alpha\frac{\omega}{\mu_0\gamma}\right) - \frac{\omega^2}{(\mu_0\gamma)^2}}e^{-i\left(kx-\frac{\pi}{2}\right)}
$$

Note that while the absolute value of $\chi_{zx}^{\mathrm{DE}}$ is constant for all wave numbers, the absolute value of $\chi_{zz}^{\mathrm{DE}}$ increases with wave number. For a better comparison of the two susceptibilities in the DE geometry, the absolute values of them are plotted in fig. A.1. As $\chi_{zx}^{\mathrm{DE}}$ is at least one order of magnitude smaller than $\chi_{zz}^{\mathrm{DE}}$ for the wave numbers observed in this thesis, only $\chi_{zx}^{\mathrm{DE}}$ is important and $\chi_{zz}^{\mathrm{DD}}$ can be neglected.

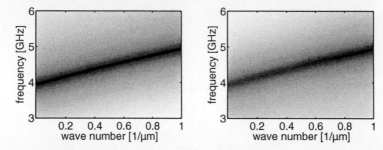

Fig. A.1: $\chi_{zx}^{\text{DE}}$ (left) and $\chi_{zz}^{\text{DE}}$ (right). The absolute value of the latter increases with increasing wave number, while it remains constant for the first one. Note that the two plots do not scale with each other; $\chi_{zx}^{\text{DE}}$ is at least one magnitude larger than $\chi_{zz}^{\text{DE}}$. Thus only $\chi_{zx}^{\text{DE}}$ is important.

### A.3.2 MSBV Geometry

The MSBV case has a different external field

$$
\mathbf{H}_0^{\text{MSBV}} = \begin{pmatrix} H_0 + h_x e^{-i\omega t} \\ 0 \\ h_z e^{-i\omega t} \end{pmatrix} \tag{A.11}
$$

which leads to a different magnetization configuration:

$$
\mathbf{M}^{\text{MSBV}} = \begin{pmatrix} M_s \\ m_y e^{i(kx-\omega t)} \\ m_z e^{i\left(kx+\frac{\pi}{2}-\omega t\right)} \end{pmatrix} \tag{A.12}
$$

The deviations from the saturation magnetization in the x-direction are very small and can be neglected. The exchange field is

$$
\mathbf{H}_{\text{exch}}^{\text{MSBV}} = \frac{2A}{\mu_0 M_s^2} \nabla^2 \mathbf{M} = \frac{2A}{\mu_0 M_s^2} \begin{pmatrix} 0 \\ \frac{\partial^2 m_y e^{i(kx-\omega t)}}{\partial x^2} \\ \frac{\partial^2 m_z e^{i\left(kx+\frac{\pi}{2}-\omega t\right)}}{\partial x^2} \end{pmatrix} . \tag{A.13}
$$

According to equation 4.1 the demagnetizing field in the MSBV case can be written as

$$
\mathbf{H}_{\text{d}}^{\text{MSBV}} = - \begin{pmatrix} 0 \\ 0 \\ m_z e^{i\left(kx+\frac{\pi}{2}-\omega t\right)} \frac{1-e^{-kd}}{kd} \end{pmatrix} \tag{A.14}
$$

Doing the analogue calculations as in the DE-case, one obtains

$$\chi_{zz}^{\mathrm{MSBV}} = \frac{m_z}{h_z} = \tag{A.15}$$

$$= \frac{M_s \left( \frac{2Ak^2}{\mu_0 M_s} + H_0 - i\alpha \frac{\omega}{\mu_0 \gamma} \right)}{\left( \frac{2Ak^2}{\mu_0 M_s} + M_s \frac{1-e^{-kd}}{kd} + H_0 - i\alpha \frac{\omega}{\mu_0 \gamma} \right) \left( \frac{2Ak^2}{\mu_0 M_s} + H_0 - i\alpha \frac{\omega}{\mu_0 \gamma} \right) - \frac{\omega^2}{(\mu_0 \gamma)^2}} e^{-i\left( kx + \frac{\pi}{2} \right)}$$

The absolute value of this susceptibility is plotted in fig. A.2. As no oscillation can be excited with the $h_x$ field in this geometry, the respective susceptibility $\chi_{zx}^{\mathrm{MSBV}}$ is zero.

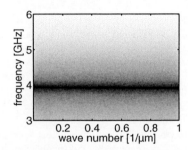

Fig. A.2: Absolute value of $\chi_{zz}^{\mathrm{MSBV}}$. Note that compared to the susceptibility for the DE geometry, the slope is much smaller here and even has a negative sign.

# Appendix B

# Magneto Optic Kerr Effect

The Magneto Optic Kerr Effect (MOKE) [Arg1955] was discovered in 1877 by John Kerr on examining light reflected from a polished electromagnet pole [Ker1876, Ker1877, Qiu2000]. The Kerr effect describes the rotation of the polarization of light upon the reflection from magnetic materials. Similarly, the Faraday effect [Far1846-1, Far1846-2] discovered some years before describes the rotation of polarization of light upon the transmission in magnetic media. Today the Kerr effect is one of the most widely used physical phenomena for the observation of magnetic properties.

Generally, the Kerr effect is a result of the interaction of the sample's internal magnetic field with the electromagnetic field of the incident light [Hub2000]. The physical principle can be explained in a very simple picture: The electrical dipole field of the incident light excited linear oscillations of the electrons inside the magnetic material. The moving charges now interact with the magnetic field, and, due to the Lorentz force, are deviated from a straight line and forced onto elliptic paths. These oscillating charges, in turn, radiate the reflected beam of light which is now no longer linearly, but elliptically polarized. However, the ellipticity of this reflected beam is so small that it is hard to detect and one merely observes a linearly polarized light with the plane of polarization slightly rotated compared to the beam of incident light.

There are four geometries that have to be taken into consideration when making use of the Kerr effect, depicted in fig. B.1, all of them being sensitive to different components of the magnetization. However, as perpendicular incidence of the probe beam is used in the setup, only the polar Kerr effect is used, and thus the setup is sensitive to the out-of plane component of the magnetization only. Disturbances from the magnetization's in-plane components are thus avoided.

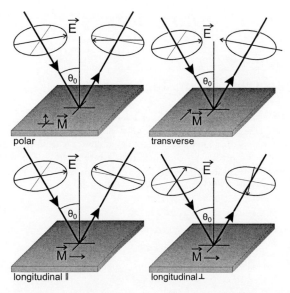

Fig. B.1: Geometries of the Magneto Optic Kerr Effect and their effect on the polarization of reflected light. The intensity of the effect is proportional to $\sin\theta_0$ for all geometries except the polar Kerr effect, where the intensity is proportional to $\cos\theta_0$. For perpendicular incidence ($\theta_0 = 0$) we only use the polar Kerr effect and are thus sensitive to the out-of-plain component of the magnetization only.

# Appendix C

# Sample Preparation

## C.1 Electron Beam Lithography

Electron Beam Lithography (EBL) is a tool widely used for the patterning of
fine structures. Due to the smaller wavelength of electrons compared to light, the
structures can be notably smaller. Moreover, in contrast to optical lithography,
no masks have to be produced, thus it is also a very flexible tool. The EBL for
the samples reported on in this thesis have all been produced using the Scanning
Electron Microscope (SEM) in Prof. Weiss' clean room facility. A recipe for
EBM shall be given in the following:

- Spin on 2 layers of PMMA to create a undercut structure in the developed
  sample:
    - Clean sample in acetone and propanol, put it into centrifuge, dry with
      nitrogen there
    - Cover sample with PMMA 200K/3.5% resist
    - Spin sample at 3000 rpm for 5 s, acceleration 0, then at 8000 rpm for
      30 s, acceleration 6
    - Remove sample from centrifuge and bake at 150° C for 7 minutes
    - Put sample into centrifuge again, cover with PMMA 950K/2.5% resist
    - Spin sample at 3000 rpm for 5 s, acceleration 0, then at 8000 rpm for
      30 s, acceleration 6
    - Remove sample from centrifuge and bake at 150° C for 7 minutes
    - Clean centrifuge with acetone, lid with propanol
- Put sample in SEM, expose properly to electron beam

- Develop Sample:

  - Put sample for 8 s into a mixture of 3 ml ethylene glycol monoethyl ether (ETX) and 7 ml methanol
  - Put sample for 8 s into methanol
  - Put sample for 30 s into propanol
  - Dry with nitrogen

Afterwards, the sample should be checked in an optical microscope and can be used for thin film deposition.

## C.2   Optical Lithography

Optical Lithography is somewhat less complicated than electron beam lithography and can be done a lot faster, especially for large structures. However, it is not as flexible and although UV light is used for exposure, structures are generally larger. For all optical lithography, the mask aligner in Prof. Weiss' clean room facility has been used. A recipe for optical lithography using the AZ5214E image reversal resist by Clariant shall be given in the following:

- Spin on image reversal resist AZ5214E:

  - Clean Sample in acetone and propanol, put it into centrifuge, dry with nitrogen there
  - Cover sample with AZ5214E image reversal resist
  - Spin sample at 3000 rpm for 40 s, acceleration 5; resist will be $\approx 1.6$ $\mu$m thick
  - Remove sample from centrifuge and bake at 90° C for 60 s

- Put sample and mask in mask aligner, expose to a dose of 55 mJ (time must be calculated from calibration chart in logbook)

- Bake at exactly 120° C for 90 s (inverse bake)

- Put sample without mask in mask aligner, expose to a dose of 300 mJ (time must be calculated from calibration chart in logbook)

- Develop sample in a mixture of 8 ml NaOH developer and 40 ml clean water for 80 s

- Stop in water for 30 s

Afterwards the sample should be checked in a microscope and can be used for thin film deposition.

## C.3   Thin Film Evaporation

The thin film evaporation for the samples reported on in this thesis has been done with the UNIVEX thin film evaporation machine in Prof. Weiss' clean room facility. $Ni_{80}Fe_{20}$ , Al and Au are evaporated using thermal evaporation, while Ti is being evaporated using e-beam evaporation. It is also possible to use Ar sputtering before thin film deposition to remove the last remnants of resist. Samples have normally been sputtered for 3-5 s at a voltage of 2.2 kV and a current of 20 mA, and for 1-2 s only when there already was another film deposited on the substrate.

## C.4   Lift-Off

After evaporation, the sample is put into acetone overnight. The acetone dissolves the resist and the thin film is being removed from the sample where the resist was, leaving only those areas covered with the thin film that were uncovered during development. Sometimes it is also necessary to boil the acetone at 60° C or to use a ultrasonic bath to remove all remnants of the resist.

# Appendix D

# SI and cgs units

The formulas in this thesis are given in SI units. However, cgs units are still widely used in magnetism. The following table gives the conversion factors from cgs to SI units.

| Magnetic Field | **H** | 1 Oe | = | $1/4\pi$ kA/m |
|---|---|---|---|---|
| Magnetic Induction | **B** | 1 G | = | 100 $\mu$T |
| Magnetization | **M** | 1 Oe | = | 1 kA/m |
| Magnetization | **M** | 1 emu/cm$^3$ | = | 1 kA/m |
| Exchange Constant | A | 1 erg/cm | = | $10^{-5}$ J/m |
| Gyromagnetic Ratio | $\gamma$ | -17.5 MHz/Oe | = | -175 GHz/T |

Table D.1: Conversion of SI and cgs units. Magnetic field and induction are equivalent in free space in the cgs system, thus they are often used interchangeably. 1 G is essentially the same as 1 Oe.

# Bibliography

[Acr2000]     ACREMAN, Y.; BACK, C.H.; BUESS, M.; PORTMANN, O.;
              VATERLAUS, A.; PESCIA, D.; AND MELCHIOR, H.:
              Imaging Precessional Motion of the Magnetization Vector
              *Science* **290**, 492 (2000)

[Acr2001]     ACREMAN, Y.; BUESS, M.; BACK, C.H.; DUMM, M.; BAYREUTHER, G.;
              AND PESCIA, D.:
              Ultrafast Generation of Magnetic Fields in a Schottky Diode
              *Nature* **414**, 6859 (2001)

[Arg1955]     ARGYRES, P.N.:
              Theory of the Faraday and Kerr Effects in Ferromagnetics
              *Physical Review* **97**, 334 (1955)

[Ari1999]     ARIAS, R.; AND MILLS, D.L.:
              Extrinsic Contributions to the Ferromagnetic Resonance Response of Ultra-
              thin Films
              *Physical Review B* **60**, 7395 (1999)

[Ari2000]     ARIAS, R.; AND MILLS, D.L.:
              Extrinsic Contributions to the Ferromagnetic Resonance Response of Ultra-
              thin Films
              *Journal of Applied Physics* **87**, 5455 (2000)

[Bai2001]     BAILLEUIL, M.; OLLIGS, D.; FERMON, C.; AND DEMOKRITOV, S.O.:
              Spin Waves Propagation and Confinement in Conducting Films at the Mi-
              crometer Scale
              *Europhysics Letters* **56**, 741 (2001)

[Bai2002]     BAILLEUIL, M.:
              Propagation et Confinement d'Ondes de Spin dans les Microstructure Mag-
              netiques
              *Thèse de Doctorat de l'cole Polytechnique (PhD thesis)*
              Service de Physique de l'tat Condens, CEA Saclay, France 2002

[Bai2003-1]   BAILLEUIL, M.; OLLIGS, D.; AND FERMON, C.:
              Propagating Spin Wave Spectroscopy in a Permalloy Film: A Quantitative
              Analysis
              *Applied Physics Letters* **83**, 972 (2003)

[Bai2003-2]     BAILLEUIL, M.; OLLIGS, D.; AND FERMON, C.:
                Micromagnetic Phase Transitions and Spin Wave Excitations in a Ferromagnetic Stripe
                *Physical Review Letters* **91**, 137204 (2003)

[Bar2003]       BARMAN, A.; KRUGLYAK, V.V.; HICKEN, R.J.; KUNDROTAITE, A.;
                AND RAHMAN, M.:
                Anisotropy, Damping, and Coherence of Magnetization Dynamics in a 10 $\mu$m
                Square $Ni_{81}Fe_{19}$ Element
                *Applied Physics Letters* **82**, 3065 (2003)

[Bar2004]       BARMAN, A.; KRUGLYAK, V.V.; HICKEN, R.J.; ROWE, J.M.;
                KUNDROTAITE, A.; SCOTT, J.; AND RAHMAN, M.:
                Imaging the Dephasing of Spin Wave Modes in a Square Thin Film Magnetic
                Element
                *Physical Review B* **69**, 174426 (2004)

[Bay2003]       BAYER, C.; DEMOKRITOV, S.O.; AND HILLEBRANDS, B.:
                Spin-Wave Wells with Multiple States Created in Small Magnetic Elements
                *Applied Physics Letters* **82**, 607 (2003)

[Bay2004]       BAYER, C.; PARK, J.P.; WANG, H.; YAN, M.; CAMPBELL, C.E.;
                AND CROWELL, P.A.:
                Spin Waves in an Inhomogeneously Magnetized Stripe
                *Physical Review B* **69**, 134401 (2004)

[Bay2005-1]     BAYER, C.; SCHULTHEISS, H.; HILLEBRANDS, B.; AND STAMPS, R.L.:
                Phase Shift of Spin Waves Traveling through a 180° Bloch-Domain-Wall
                *IEEE Transactions on Magnetics* **41**, 3094 (2005)

[Bay2005-2]     BAYER, C.; JORZICK, J.; HILLEBRANDS, B.; DEMOKRITOV, S.O.;
                KOUBA, R.; BOZINOSKI, R.; SLAVIN, A.N.; GUSLIENKO, K.Y.;
                BERKOV, D.V.; GORN, N.L.; AND KOSTYLEV, M.P.:
                Spin-Wave Excitations in Finite Rectangular Elements of $Ni_{80}Fe_{20}$
                *Physical Review B* **72**, 064427 (2005)

[Bay2006]       BAYER, C.; KOSTYLEV, M.P.; AND HILLEBRANDS, B.:
                Spin-Wave Eigenmodes of an Infinite Thin Film with Periodically Modulated
                Exchange Bias Field
                *Applied Physics Letters* **88**, 112504 (2006)

[Bel2004]       BELOV, M.; LIU, Z.; SYDORA, S.D.; AND FREEMAN, M.R.:
                Modal Oscillation Control in Internally Patterned $Ni_{80}Fe_{20}$ Thin Film Microstructures
                *Physical Review B* **69**, 094414 (2004)

[Blo1930]       BLOCH, F.:
                Zur Theorie des Ferromagnetismus
                *Zeitschrift für Physik* **61**, 206 (1930)

[Bol2007]       BOLTE, M.; MEIER, G.; AND BAYER, C.:
                Symmetry Dependence of Spin-Wave Eigenmodes in Landau-Domain Patterns
                *Journal of Magnetism and Magnetic Materials* **316**, e526 (2007)

[Bra2004]  BRAUDE, V.; AND SONIN, E.B.:
Excitation of Spin Waves in Superconducting Ferromagnets
*Physical Review Letters* **93**, 117001 (2004)

[Bro1963]  BROWN, JR., W.J.:
Micromagnetics
*Interscience Tracts on Physics and Astronomy* **18** (1963)
John Wiley & Sons, New York, London

[Bue2003]  BUESS, M.; ACREMAN, Y.; KASHUBA, A.; BACK, C.H.; AND PESCIA, D.:
Pulsed Precessional Motion on the 'Back of an Envelope'
*Journal of Physics C: Condensed Matter* **15**, R1093 (2003)

[Bue2004-1]  BUESS, M.; HÖLLINGER, R.; HAUG, T.; PERZLMAIER, K.; KREY, U.;
PESCIA, D.; SCHEINFEIN, M.R.; WEISS, D.; AND BACK, C.H.:
Fourier Transform Imaging of Spin Vortex Eigenmodes
*Physical Review Letters* **93**, 077207 (2004)

[Bue2004-2]  BUESS, M.; KNOWLES, T.P.J.; RAMSPERGER, U.; PESCIA, D.;
AND BACK, C.H.:
Phase-Resolved Pulsed Precessional Motion at a Schottky Barrier
*Physical Review B* **69**, 174422 (2004)

[Bue2005-1]  BUESS, M.; KNOWLES, T.P.J.; HÖLLINGER, R.; HAUG, T.; KREY, U.;
WEISS, D.; PESCIA, D.; SCHEINFEIN, M.R.; AND BACK, C.H.:
Excitations with Negative Dispersion in a Spin Vortex
*Physical Review B* **71**, 104415 (2005)

[Bue2005-2]  BUESS, M.; HAUG, T.; SCHEINFEIN, M.R.; AND BACK, C.H.:
Micromagnetic Dissipation, Dispersion, and Mode Conversion in Thin
Permalloy Platelets
*Physical Review Letters* **94**, 127205 (2005)

[Bue2005-3]  BUESS, M.:
Pulsed Precessional Motion
*Dissertation (PhD Thesis)*
ETH Zürich, Zürich, Switzerland, and Universität Regensburg, Regensburg
2005

[Bue2006]  BUESS, M.; RAABE, J.; PERZLMAIER, K.; BACK, C.H.;
AND QUITMANN, C.:
Interaction of Magnetostatic Excitations with 90° Domain Walls in
Micrometer-Sized Permalloy Squares
*Physical Review B* **74**, 100404(R) (2006)

[Büt2000]  BÜTTNER, O.; BAUER, M.; DEMOKRITOV, S.O.; HILLEBRANDS, B.;
KIVSHAR, Y.S.; GRIMALSKY, V.; RAPOPORT, YU.; AND SLAVIN, A.N.:
Linear and Nonlinear Diffraction of Dipolar Spin Waves in Yttrium Iron Gar-
net Films Observed by Space- and Time-Resolved Brillouin Light Scattering
*Physical Review B* **61**, 11576 (2000)

[Chi1993]     CHIAO, R.Y.:
              Superluminal (But Causal) Propagation of Wave Packets in Transparent Me-
              dia With Inverted Atomic Populations
              *Physical Review A* **48**, R34 (1993)

[Cho2006]     CHOI, S.; LEE, K.-S.; AND KIM, S.-K.:
              Spin-Wave Interference
              *Applied Physics Letters* **89**, 062501 (2006)

[Cho2007]     CHOI, S.; LEE, K.-S.; GUSLIENKO, K.Y.; AND KIM, S.-K.:
              Strong Radiation of Spin Waves by Core Reversal of a Magnetic Vortex and
              Their Wave Behaviors in Magnetic Nanowire Waveguides
              *Physical Review Letters* **98**, 087205 (2007)

[Cou2004]     COUNIL, G.; KIM, J.-V.; DEVOLDER, T.; AND CHAPPERT, C.:
              Spin Wave Contributions to the High-Frequency Magnetic Response of Thin
              Films Obtained with Inductive Methods
              *Journal of Applied Physics* **95**, 5646 (2004)

[Cov2002]     COVINGTON, M.; CRAWFORD, T.M.; AND PARKER, G.J.:
              Time-Resolved Measurement of Propagating Spin Waves in Ferromagnetic
              Thin Films
              *Physical Review Letters* **89**, 237202 (2002)
              and: Erratum
              *Physical Review Letters* **92**, 089903 (2004)

[Cra2003]     CRAWFORD, T.M.; COVINGTON, M.; AND PARKER, G.J.:
              Time-Domain Excitation of Quantized Magnetostatic Spin-Wave Modes in
              Patterned NiFe Thin Film Ensembles
              *Physical Review B* **67**, 024411 (2003)

[Dam1960]     DAMON, R.W.; AND ESHBACH, J.R.:
              Magnetostatic Modes of a Ferromagnetic Slab
              *Journal of Applied Physics* **11**, 104S (1960)

[Dam1961]     DAMON, R.W.; AND ESHBACH, J.R.:
              Magnetostatic Modes of a Ferromagnet Slab
              *Journal of Physics and Chemistry of Solids* **19**, 308 (1961)

[Dem2001]     DEMOKRITOV, S.O.; HILLEBRANDS, B.; AND SLAVIN, A.N.:
              Brillouin Light Scattering Studies of Confined Spin Waves: Linear and Non-
              linear Confinement
              *Physics Reports* **348**, 441 (2001)

[Dem2003 1]   DEMOKRITOV, S.O.; SERGA, A.A.; DEMIDOV, V.E.; HILLEBRANDS, B.;
              KOSTYLEV, M.P.; AND KALINIKOS, B.A.:
              Experimental Observation of Symmetry-Breaking Nonlinear Modes in an Ac-
              tive Ring
              *Nature* **426**, 159 (2003)

[Dem2003 2]   DEMOKRITOV, S.O.:
              Dynamic Eigen-Modes in Magnetic Stripes and Dots
              *Journal of Physics: Condensed Matter* **15**, S2575 (2003)

[Dem2004] DEMOKRITOV, S.O.; SERGA, A.A.; ANDRÉ, A.; DEMIDOV, V.E.; KOSTYLEV, M.P.; AND HILLEBRANDS, B.:
Tunneling of Dipolar Spin Waves through a Region of Inhomogeneous Magnetic Field
*Physical Review Letters* **93**, 047201 (2004)

[Dem2006] DEMOKRITOV, S.O.; DEMIDOV, V.E.; DZYAPKO, O.; MELKOV, G.A.; SERGA, A.A.; HILLEBRANDS, B.; AND SLAVIN, A.N.:
Bose-Einstein Condensation of Quasi-Equilibrium Magnons at Room Temperature Under Pumping
*Nature* **443**, 430 (2006)

[Dem2007] DEMIDOV, V.E.; HANSEN, U.-H.; AND DEMOKRITOV, S.O.:
Spin-Wave Eigenmodes of a Saturated Magnetic Square at Different Precession Angles
*Physical Review Letters* **98**, 157203 (2007)

[Di1960] DILLON JR., J.F.:
Magnetostatic Modes in Disks and Rods
*Journal of Applied Physics* **31**, 1605 (1960)

[Dör1948] DÖRING, W.:
Über die Trägheit der Wände zwischen Weißschen Bezirken
*Zeitschrift für Naturforschung* **3a**, 373 (1948)

[Ele1996] ELEZZABI, A.Y.; AND FREEMAN, M.R.:
Ultrafast Magneto-Optic Sampling of Picosecond Current Pulses
*Physical Review Letters* **93**, 047201 (2004)

[Esh1962] ESHBACH, J.R.:
Spin-Wave Propagation and the Magnetoelastic Interaction in Yttrium Iron Garnet
*Physical Review Letters* **8**, 357 (1962)

[Esh2007] ESCHAGHIAN-WILNER, M.M.; KHITUN, A.; NAVAB, S.; AND WANG, K.L.:
A Nano-Scale Architecture for Constant Time Image Processing
*Physica Status Solidi (A)* **204**, 1931 (2007)

[Far1846-1] FARADAY, M.:
Experimental Researches in Electricity
*Proceedings of the Royal Society of London* **5**, 592 (1846)

[Far1846-2] FARADAY, M.:
Experimental Researches in Electricity - Nineteenth Series
*Philosophical Transactions of the Royal Society of London* **136**, 1 (1846)

[Fel1960] FELDTKELLER, E.:
Die Ummagnetisierung Anisotroper Nickeleisenschichten in der Schweren Richtung - II. Hysterese und Domänenverhalten
*Zeitschrift für angewandte Physik* **13**, 161 (1961)

[Fuc1960] FUCHS, E.:
Die Ummagnetisierung Dünner Nickeleisenschichten in der Schweren Richtung
*Zeitschrift für Angewandte Physik* **13**, 157 (1961)

[Fle1960]      FLETCHER, P.C.; AND KITTEL, C.:
               Considerations on the Propagation and Generation of Magnetostatic Waves
               and Spin Waves
               *Physical Review* **120**, 2004 (1960)

[Ger2001]      GERRITS, TH.; VAN DEN BERG, H.A.M.; HOHLFELD, J.; GIELKENS, O.;
               VEENSTRA, K.J.; BAL, K.; AND RASING, TH.:
               Precession Dynamics in NiFe Thin Films, Induced by Short Magnetic In-
               Plane Field Pulses Generated by a Photoinductive Switch
               *Journal of the Magnetics Society of Japan* **25**, 192 (2001)

[Ger2007]      GERRITS, TH.; KRIVOSIK, P.; SCHNEIDER, M.L.; PATTON, C.E.;
               AND SILVA, T.J.:
               Direct Detection of Nonlinear Ferromagnetic Resonance in Thin Films by
               the Magneto-Optical Kerr Effect
               *Physical Review Letters* **98**, 207602 (2007)

[Gie2005]      GIESEN, F.; PODBIELSKI, J.; KORN, T.; STEINER, M.; VAN STAA, A.;
               AND GRUNDLER, D.:
               Hysteresis and Control of Ferromagnetic Resonances in Rings
               *Applied Physics Letters* **86**, 112510 (2005)

[Gie2007]      GIESEN, F.; PODBIELSKI, J.; AND GRUNDLER, D.:
               Mode Localization Transition in Ferromagnetic Microscopic Rings
               *Physical Review B* **76**, 014431 (2007)

[Gil1955]      GILBERT, T.L.:
               A Langrangian Formulation of the Gyromagnetic Equation of the Magneti-
               zation Field
               *Physical Review* **100**, 1243 (1955)

[Gio2004]      GIOVANNINI, L.; MONTONCELLO, F.; NIZZOLI, F.; GUBBIOTTI, G.;
               CARLOTTI, G.; OKUNO, T.; SHINJO, T.; AND GRIMSDITCH, M.:
               Spin Excitations of Nanometric Cylindrical Dots in Vortex and Saturated
               Magnetic States
               *Physical Review B* **70**, 172404 (2004)

[Gub2003]      GUBBIOTTI, G.; CARLOTTI, G.; OKUNO, T.; SHINJO, T.; NIZZOLI, F.;
               AND ZIVERI, R.:
               Brillouin Light Scattering Investigation of Dynamic Spin Modes Confined in
               Cylindrical Permalloy Dots
               *Physical Review B* **68**, 184409 (2003)

[Gub2005]      GUBBIOTTI, G.; CARLOTTI, G.; OKUNO, T.; GRIMSDITCH, M.;
               GIOVANNINI, L.; MONTONCELLO, F.; AND NIZZOLI, F.:
               Spin Dynamics in Thin Nanometric Elliptical Permalloy Dots: A Brillouin
               Light Scattering Investigation as a Function of Dot Eccentricity
               *Physical Review B* **72**, 184419 (2005)

[Gus2003]      GUSLIENKO, K.Y.; CHANTRELL, R.W.; AND SLAVIN, A.N.:
               Dipolar Localization of Quantized Spin-Wave Modes in Thin Rectangular
               Magnetic Elements
               *Physical Review B* **68**, 024422 (2003)

[Har1968]     HARTE, K.J.:
              Theory of Magnetization Ripple in Ferromagnetic Films
              *Journal of Applied Physics* **39**, 1503 (1968)

[Her1951]     HERRING, C.; AND KITTEL, C.:
              On the Theory of Spin Waves in Ferromagnetic Media
              *Physical Review* **81**, 869 (1951)

[Her2004]     HERTEL, R.; WULFHEKEL, W.; AND KIRSCHNER, J.:
              Domain-Wall Induced Phase Shifts in Spin Waves
              *Physical Review Letters* **93**, 257202 (2004)

[Hie1997]     HIEBERT, W.K.; STANKIEWICZ, A.; AND FREEMAN, M.R.:
              Direct Observation of Magnetic Relaxation in a Small Permalloy Disk by
              Time-Resolved Scanning Kerr Microscopy
              *Physical Review Letters* **79**, 1134 (1997)

[Hil1990]     HILLEBRANDS, B.:
              Spin-Wave Calculations for Multilayered Structures
              *Physical Review B* **41**, 530 (1990)

[Ho2004-1]    HO, J.; KHANNA, F.C.; AND CHOI, B.C.:
              Radiation-Spin Interaction, Gilbert Damping, and Spin Torque
              *Physical Review Letters* **92**, 097601 (2004)

[Ho2004-2]    HO, J.; KHANNA, F.C.; AND CHOI, B.C.:
              Combination of Dynamical Invariant Method and Radiation-Spin Interaction
              to Calculate Magnetization Damping
              *Physical Review B* **70**, 172402 (2004)

[Hof2007]     HOFFMANN, F.; WOLTERSDORF, G.; PERZLMAIER, K.; SLAVIN, A.N.;
              TIBERKEVICH, V.S.; BISCHOF, A.; WEISS, D.; AND BACK, C.H.:
              Mode Degeneracy due to Vortex Core Removal in Magnetic Disks
              *Physical Review B* **76**, 014416 (2007)

[Höl2002]     HÖLLINGER, R.; KILLINGER, A.; AND KREY, U.:
              Statistics and Fast Dynamics of Nanomagnets with Vortex Structure
              *Journal of Magnetism and Magnetic Materials* **261**, 178 (2003)

[Hub2000]     HUBERT, A.; SCHÄFER, R.:
              Magnetic Domains - The Analysis of Magnetic Microstructures
              Corrected Printing 2000
              Springer-Verlag Berlin, Heidelberg

[Iva2002]     IVANOV, B.A.; AND ZASPEL, C.E.:
              Magnon Modes for Thin Circular Vortex-State Magnetic Dots
              *Applied Physics Letters* **81**, 1261 (2002)

[Iva2005]     IVANOV, B.A.; AND ZASPEL, C.E.:
              High Frequency Modes in Vortex-State Nanomagnets
              *Physical Review Letters* **94**, 027205 (2005)

[Jor1999-1]    JORZICK, J.; DEMOKRITOV, S.O.; MATHIEU, C.; HILLEBRANDS, B.;
               BARTENLIAN, B; CHAPPERT, C.; ROUSSEAUX, F.; AND SLAVIN, A.N.:
               Brillouin Light Scattering from Quantzized Spin Waves in Micron-Sized Mag-
               netic Wires
               *Physical Review B* **60**, 15194 (1999)

[Jor1999-2]    JORZICK, J.; DEMOKRITOV, S.O.; HILLEBRANDS, B.; BARTENLIAN, B;
               CHAPPERT, C.; DECANINI, D.; ROUSSEAUX, F.; AND CAMBRIL, E.:
               Spin-Wave Quantization and Dynamic Coupling in Micron-Sized Circular
               Magnetic Dots
               *Applied Physics Letters* **75**, 3859 (1999)

[Jor2001]      JORZICK, J.; KRÄMER, C.; DEMOKRITOV, S.O.; HILLEBRANDS, B.;
               BARTENLIAN, B; CHAPPERT, C.; DECANINI, D.; ROUSSEAUX, F.; CAM-
               BRIL, E.; SONDERGARD, E.; BAILLEUIL, M.; AND FERMON, C.:
               Spin Wave Quantization in Laterally Confined Magnetic Structures
               *Journal of Applied Physics* **89**, 7091 (2001)

[Jor2002]      JORZICK, J.; DEMOKRITOV, S.O.; HILLEBRANDS, B.; BAILLEUIL, M.;
               FERMON, C.; GUSLIENKO, K.Y.; SLAVIN, A.N.; BERKOV, D.V.; AND
               GORN, N.L.:
               Spin Wave Wells in Nonellipsoidal Micrometer Size Magnetic Elements
               *Physical Review Letters* **88**, 047204 (2002)

[Kak2004]      KAKAZEI, G.N.; WIGEN, P.E.; GUSLIENKO, K.YU.; NOVOSAD, V.;
               SLAVIN, A.N.; GOLUB, V.O.; LESNIK, N.A.; AND OTANI, Y.:
               Spin-Wave Spectra of Perpendicularly Magnetized Circular Submicron Dot
               Arrays
               *Applied Physics Letters* **85**, 443 (2004)

[Kal1986]      KALINIKOS, B.A., AND SLAVIN, A.N.:
               Theory of Dipole-Exchange Spin Wave Spectrum for Ferromagnetic Films
               with Mixed Exchange Boundary Conditions
               *Journal of Physics and Chemistry of Solids* **19**, 7013 (1986)

[Ker1876]      KERR, J.:
               On the Magnetism of Light and the Illumination of Magnetic Lines of Force
               *Report of the British Association for the Advancement of Science* **S5**, 85
               (1876)

[Ker1877]      KERR, J.:
               A New Relation between Electricity and Light
               *Philosophical Magazine* **3**, 321 (1877)

[Kim2005]      KIMEL, A.V.; KIRILYUK, A.; USACHEV, P.A.; PISAREV, R.V.;
               BALBASHOV, A.M.; AND RASING, TH:
               Ultrafast Non-Thermal Control of Magnetization by Instantaneous Photo-
               magnetic Pulses
               *Nature* **435**, 655 (2005)

[Kit1948]      KITTEL, C.:
               On the Theory of Ferromagnetic Resonance Absorption
               *Physical Review* **73**, 155 (1948)

[Kit1949]     KITTEL, C.:
              Physical Theory of Ferromagnetic Domains
              *Reviews of Modern Physics* **21**, 541 (1949)

[Kos2007]     KOSTYLEV, M.P.; GUBBIOTTI, G.; HU, J.-G.; CARLOTTI, G.; ONO, T.;
              AND STAMPS, R.L.:
              Dipole-Exchange Propagating Spin-Wave Modes in Metallic Ferromagnetic
              Stripes
              *Physical Review B* **76**, 054422 (2007)

[Lan1935]     LANDAU, L.; AND LIFSHITZ, E.:
              On the Theory of the Dispersion of Magnetic Permeability in Ferromagnetic
              Bodies
              *Physikalische Zeitschrift der Sowjetunion* **8**, 153 (1935)

[LeC1958]     LECRAW, R.C.; SPENCER, E.G.; AND PORTER, C.S.:
              Ferromagnetic Resonance Line Width in Yttrium Iron Garnet Single Crystals
              *Physical Review* **110**, 1311 (1958)

[Liu2007]     LIU, Z.; GIESEN, F.; ZHU, X.; SYDORA, R.D.; AND FREEMAN, F.R.:
              Spin Wave Dynamics and the Determination of Intrinsic Damping in Locally
              Extended Permalloy Thin Films
              *Physical Review Letters* **98**, 087201 (2007)

[Lot2004]     LOTTERMOSER, T.; LONKAI, T.; AMANN, U.; HOHLWEIN, D.; IHRINGER,
              J.; AND FIEBIG, M.:
              Magnetic Phase Control by an Electric Field
              *Nature* **430**, 541 (2004)

[McM1998]     MCMICHAEL, R.D.; STILES, M.D.; CHEN, P.J.;
              AND EGELHOFF JR., W.F.:
              Ferromagnetic Resonance Linewidth in Thin Films Coupled to NiO
              *Journal of Applied Physics* **83**, 7037 (1998)

[Mil2002]     MILTAT, J.; ALBUQUERQUE, G.; AND THIAVILLE, A.:
              An Introduction to Micromagnetics in the Dynamic Regime
              in:
              HILLEBRANDS, B.; AND OUNADJELA, K. (EDS.):
              Spin Dynamics in Confined Magnetic Structures I
              *Topics in Applied Physics* **83**, 1-33 (2002)
              Springer-Verlag Berlin, Heidelberg

[Mil2003]     MILLS, D.L.; AND REZENDE, S.M.:
              Spin Damping in Ultrathin Magnetic Films
              in:
              HILLEBRANDS, B.; AND OUNADJELA, K. (EDS.):
              Spin Dynamics in Confined Magnetic Structures II
              *Topics in Applied Physics* **87**, 27-58 (2003)
              Springer-Verlag Berlin, Heidelberg

[Neu2006-1]  NEUDECKER, I.; KLÄUI, M.; PERZLMAIER, K.; BACKES, D.;
HEYDERMAN, L.J.; VAZ, C.A.F.; BLAND, J.A.C.; RÜDIGER, U.;
AND BACK, C.H.:
Spatially Resolved Dynamic Eigenmode Spectrum of Co Rings
*Physical Review B* **73**, 134426 (2006)

[Neu2006-2]  NEUDECKER, I.; PERZLMAIER, K.; HOFFMANN, F.; WOLTERSDORF, G.;
BUESS, M.; WEISS, D.; AND BACK, C.H.:
Modal Spectrum of Permalloy Disks Excited by In-Plane Magnetic Fields
*Physical Review Letters* **96**, 057207 (2006)

[Ney2003]  NEY, A.; PAMPUCH, C.; KOCH, R.; AND PLOOG, K.H.:
Programmable Computing with a Single Magnetoresistive Element
*Nature* **425**, 485 (2003)

[Nib2003]  NIBARGER, J.P.; LOPUSNIK, R.; AND SILVA, T.J.:
Damping as a Function of Pulsed Field Amplitude and Bias Field in Thin
Film Permalloy
*Applied Physics Letters* **82**, 2112 (2003)

[Par2002]  PARK, J.P.; EAMES, P.; ENGEBRETSON, D.M.; BEREZOVSKY, J.;
AND CROWELL, P.A.:
Spatially Resolved Dynamics of Localized Spin-Wave Modes in Ferromag-
netic Wires
*Physical Review Letters* **89**, 277201 (2002)

[Par2003]  PARK, J.P.; EAMES, P.; ENGEBRETSON, D.M.; BEREZOVSKY, J.;
AND CROWELL, P.A.:
Imaging of Spin Dynamics in Closure Domain and Vortex Structures
*Physical Review B* **67**, 020403(R) (2003)

[Par2004]  PARIMI, P.V.; LU, W.T.; SOKOLOFF, J.; DEROV, J.S.; AND SRIDHAR, S.:
Negative Refraction and Left-Handed Electromagnetism in Microwave Pho-
tonic Crystals
*Physical Review Letters* **92**, 127401 (2004)

[Par2005]  PARK, J.P.; AND CROWELL, P.A.:
Interactions of Spin Waves with a Magnetic Vortex
*Physical Review Letters* **95**, 167201 (2005)

[Pen2003]  PENDRY, J.B.; AND SMITH, D.R.:
Comment on "Wave Refraction in Negative-Index Media: Always Positive
and Very Inhomogeneous"
*Physical Review Letters* **90**, 029703 (2003)

[Per2005]  PERZLMAIER, K.; BUESS, M.; BACK, C.H.; DEMIDOV, V.E.;
HILLEBRANDS, B.; AND DEMOKRITOV, S.O.:
Spin-Wave Eigenmodes of Permalloy Squares with a Closure Domain Struc-
ture
*Physical Review Letters* **94**, 057202 (2005)

[Pim2007]     PIMENOV, A.; LOIDL, A.; GEHRKE, K.; MOSHNYAGA, V.;
              AND SAMWER, K.:
              Negative Refraction Observed in a Metallic Ferromagnet in the Gigahertz
              Frequency Range
              *Physical Review Letters* **98**, 197401 (2007)

[Pri1998]     PRINZ, G.A.:
              Magnetoelectronics
              *Science* **282**, 1660 (1998)

[Qiu2000]     QIU, Z.Q.; BADER, S.D.:
              Surface Magneto-Optic Kerr Effect
              *Review of Scientific Instruments* **71**, 1243 (2000)

[Sch2004]     SCHNEIDER, M.L.; KOS, A.B.; AND SILVA, T.J.:
              Finite Coplanar Waveguide Width Effects in Pulsed Inductive Microwave
              Magnetometry
              *Applied Physics Letters* **85**, 254 (2004)

[Sch2005]     SCHNEIDER, M.L.; GERRITS, TH.; KOS, A.B.; AND SILVA, T.J.:
              Gyromagnetic Damping and the Role of Spin-Wave Generation in Pulsed
              Inductive Microwave Magnetometry
              *Applied Physics Letters* **87**, 072509 (2005)

[Sch2006]     SCHNEIDER, T.; KOSTYLEV, M.; LEVEN, B.; SERGA, A.;
              AND HILLEBRANDS, B.:
              Spin Wave Logic Gates
              *Verhandlungen der Deutschen Physikalischen Gesellschaft (VI)* **41**, 1/355
              (2006)

[Ser2003-1]   SERGA, A.A.; DEMOKRITOV, S.O.; HILLEBRANDS, B.; MIN, S.-G.;
              AND SLAVIN, A.N.:
              Phase Control of Nonadiabatic Parametric Amplification of Spin Wave Pack-
              ets
              *Journal of Applied Physics* **93**, 8585 (2003)

[Ser2003-2]   SERGA, A.A.; DEMOKRITOV, S.O.; HILLEBRANDS, B.; AND SLAVIN, A.N.:
              Formation of Envelope Solitons from Parametrically Amplified and Conju-
              gated Spin Wave Pulses
              *Journal of Applied Physics* **93**, 8758 (2003)

[Ser2003-3]   SERGA, A.A.; KOSTYLEV, M.P.; KALINIKOS, B.A.; DEMOKRITOV, S.O.;
              HILLEBRANDS, B.; AND BENNER, H.:
              Parametric generation of Solitonlike Spin-Wave Pulses in Ring Resonators
              Based on Ferromagnetic Films
              *Journal of Experimental and Theoretical Physics Letters* **77**, 300 (2003)

[Ser2004]     SERGA, A.A.; DEMOKRITOV, S.O.; AND HILLEBRANDS, B.:
              Self-Generation of Two-Dimensional Spin-Wave Bullets
              *Physical Review Letters* **92**, 117203 (2004)

[Sil2002]     SILVA, T.J.; PUFALL, M.R.; AND KABOS, P.:
              Nonlinear Magneto-Optic Measurement of Flux Propagation Dynamics in
              Thin Permalloy Films
              *Journal of Applied Physics* **91**, 1066 (2002)

[Sla2003]     SLAVIN, A.N.; BÜTTNER, O.; BAUER, M.; DEMOKRITOV, S.O.; HILLE-
              BRANDS, B.; KOSTYLEV, M.P.; KALINIKOS, B.A.; GRIMALSKY, V.V.;
              AND RAPOPORT, YU.:
              Collision Properties of Quasi-One-Dimensional Spin Wave Solitons and Two-
              Dimensional Spin Wave Bullets
              *Chaos* **13**, 692 (2003)

[Spa1970]     SPARKS, M.:
              Magnetostatic Modes in an Infinite Circular Disk
              *Solid State Communications* **8**, 731 (1970)

[Sta2007-1]   STANCIU, C.D.; HANSTEEN, F.; KIMEL, A.V.; TSUKAMOTO, A.;
              ITOH, A.; KIRILYUK, A.; AND RASING, TH.:
              Ultrafast Interaction of the Angular Momentum of Photons with Spins in
              the Metallic Amorphous Alloy GdFeCo
              *Physical Review Letters* **98**, 207401 (2007)

[Sta2007-2]   STANCIU, C.D.; HANSTEEN, F.; KIMEL, A.V.; KIRILYUK, A.;
              TSUKAMOTO, A.; ITOH, A.; AND RASING, TH.:
              All-Optical Magnetic Recording with Circularly Polarized Light
              *Physical Review Letters* **99**, 047601 (2007)

[Sto1948]     STONER, E.C., AND WOHLFAHRT, E.P.:
              A Mechanism of Magnetic Hysteresis in Heterogeneous Alloys
              *Philosophical Transactions of the Royal Society of London, Series A* **240**, 74
              (1948)

[Sto2004]     STOLL, H; PUZIC, A.; VAN WAEYENBERGE, B.; FISCHER, P.; RAABE, J.;
              BUESS, M.; HAUG, T.; HÖLLINGER, R.; BACK, C.H.; WEISS, D.;
              AND DENBEAUX, G.:
              High-Resolution Imaging of Fast Magnetization Dynamics in Magnetic
              Nanostructures
              *Applied Physics Letters* **84**, 3328 (2004)

[Suh1957]     SUHL, H.:
              The Theory of Ferromagnetic Resonance at High Signal Powers
              *Journal of Physics and Chemistry of Solids* **1**, 209 (1957)

[Tam2002]     TAMARU, S.; BAIN, J.A.; VAN DE VEERDONK, R.J.M.; CRAWFORD, T.M.;
              COVINGTON, M.; AND KRYDER, M.H.:
              Imaging of Quantized Magnetostatic Modes Using Spatially Resolved Ferro-
              magnetic Resonance
              *Journal of Applied Physics* **91**, 8034 (2002)

[Tam2004]     TAMARU, S.; BAIN, J.A.; VAN DE VEERDONK, R.J.M.; CRAWFORD, T.M.;
              COVINGTON, M.; AND KRYDER, M.H.:
              Measurement of Magnetostatic Mode Excitation and Relaxation in Permalloy
              Films Using Scanning Kerr Microscopy
              *Physical Review B* **70**, 104416 (2004)

[Teh1999]    TEHRANI, S.; CHEN, E.; DURLAM, M.; DEHERRERA, M.;
             SLAUGHTER, J.M.; SHI, J.; AND KERSZYKOWSKI, G.:
             High Density Submicron Magnetoresistive Random Access Memory
             *Journal of Applied Physics* **85**, 5822 (1999)

[Vee2006]    VEERAKUMAR, V.; AND CAMLEY, R.E.:
             Magnon Focusing in Thin Ferromagnetic Films
             *Physical Review B* **74**, 214401 (2006)

[Vas2007]    VASILIEV, S.V.; KRUGLYAK, V.V.; SOKOLOVSKII, M.L.;
             AND KUCHKO, A.N.:
             Spin Wave Interferometer Employing a Local Nonuniformity of the Effective
             Magnetic Field
             *Journal of Applied Physics* **101**, 113919 (2007)

[vKa2006]    VAN KAMPEN, M.; JOZSA, C.; KOHLHEPP, J.T.; LECLAIR, P.;
             LAGAE, L.; DE JONGE, W.J.M.; AND KOOPMANS, B.:
             All-Optical Probe of Coherent Spin Waves
             *Physical Review Letters* **88**, 227201 (2002)

[Wei2007]    WEISHEIT, M.; FÄHLER, S.; MARTY, A.; SOUCHE, Y.; POINSIGNON, C.;
             AND GIVORD, D.:
             Electric Field-Induced Modification of Magnetism in Thin-Film Ferromag-
             nets
             *Science* **315**, 349 (2007)

[Whi2007]    WHITE, R.M.:
             Quantum Theory of Magnetism - Magnetic Properties of Materials
             *Springer Series in Solid-State Sciences* **32** (2007)
             Third Edition
             Springer-Verlag Berlin, Heidelberg

[Wol2001]    WOLF, S.A.; AWSCHALOM, D.D.; BUHRMAN, D.A.; DAUGHTON, J.M.;
             VON MOLNÁR, S.; ROUKES, M.L.; CHTCHELKANOVA, A.Y.;
             AND TREGER, D.M.:
             Spintronics: A Spin-Based Electronics Vision for the Future
             *Science* **294**, 1488 (2001)

[Wol2004]    WOLTERSDORF, G.:
             Spin-Pumping and Two-Magnon Scattering in Magnetic Multilayers
             *PhD Thesis, Simon Fraser University,* Burnaby, Canada (2004)

[Wu2004]     WU, M.; KALINIKOS, B.A.; AND PATTON, C.E.:
             Generation of Dark and Bright Spin Wave Envelope Soliton Trains through
             Self-Modulational Instability in Magnetic Films
             *Physical Review Letters* **93**, 157207 (2004)

[Wu2006]     WU, M.; KALINIKOS, B.A.; KRIVOSIK, P.; AND PATTON, C.E.:
             Fast Pulse-Excited Spin Waves in Yttrium Iron Garnet Thin Films
             *Journal of Applied Physics* **99**, 013901 (2006)

[Zhu2005-1]    ZHU, X.; MALAC, M.; LIU, Z.; QIAN, H; METLUSHKO, V.;
               AND FREEMAN, M.R.:
               Broadband Spin Dynamics of Permalloy Rings in the Circulation State
               *Applied Physics Letters* **86**, 262502 (2005)

[Zhu2005-2]    ZHU, X.; LIU, Z.; METLUSHKO, V.; GRÜTTER, P.; AND FREEMAN, M.R.:
               Broadband Spin Dynamics of the Magnetic Vortex State: Effect of the Pulsed
               Field Direction
               *Physical Review B* **71**, 180408(R) (2005)

# Appendix E

# Danksagung

Nach 3 Jahren Promotion möchte ich mich bedanken:

- ...bei Prof. Dr. Christian H. Back, der mir die Möglichkeit geboten hat, an seinem Lehrstuhl dieses interessante und sehr ergiebige Thema in einer Doktorarbeit zu behandeln;

- ...bei Prof. Dr. Dieter Weiss, in dessen Reinraum ich meine Proben herstellen durfte; sowie bei allen Mitarbeitern des Lehrstuhl Weiss, die dazu beigtragen haben, dass dort alles funktioniert hat;

- ...bei der Studienstiftung des Deutschen Volkes, die mir mit einem Stipendium meine Promotion finanziert und die Möglichkeit zur Teilnahme an vielen interessanten Veranstaltungen und Seminaren geboten hat;

- ...bei Dr. Georg Woltersdorf für sein umfangreiches Wissen auf dem Gebiet des Magnetismus das er stets zu teilen bereit war, sowie für die Zusammenarbeit im Labor;

- ...bei Prof. Dr. Uwe Krey für viele interessante Diskussionen;

- ...bei Wolfgang Scheibenzuber, Frank Hoffmann und Georg Woltersdorf für die Zusammenarbeit in einem Projekt, in dessen Rahmen die Messungen aus Kaptitel 6.3.2 sowie aus fig. 6.15 entstanden sind; bei Georg Woltersdorf auch besonders für die Herstellung der hierfür benötigten Ring Interferometer;

- ...bei Tobias Stöckl, Dieter Schierl, der Elektronik- und der Mechanikwerkstatt für die schnelle Hilfe, wenn man sie gebraucht hat;

- ...bei Georg Woltersdorf, Christian Back und Angelika Batke für's Korrekturlesen der Dissertation;

- ...bei meinen Laborgenossen Frank Hoffmann und Rüdiger Pürner für die Zusammenarbeit im Labor;

- ...bei meinen Zimmergenossen Tobias Martin, Florian Nitsch und Markus Maier für die angenehme Atmosphäre;

- ...bei den regelmässigen Mensagängern Christian Hurm, Jürgen Gründmayer, Christian Dietrich, Martin Beer, Matthias Kiessling, Matthias Neu, Roman Grothausmann, Martin Brunner und vielen anderen; gemeinsam schmeckt's halt doch besser;

- ...bei allen anderen Mitgliedern des Lehrstuhl Back für die angenehme Atmosphäre;

- ...bei meinen Eltern, meinem Bruder und Angelika für die moralische Unterstützung, die man immer mal brauchen kann.

# Appendix F

# Publications

BUESS, M.; HÖLLINGER, R.; HAUG, T.; PERZLMAIER, K.; KREY, U.; PESCIA, D.; SCHEINFEIN, M.R.; WEISS, D.; AND BACK, C.H.:
Fourier Transform Imaging of Spin Vortex Eigenmodes
*Physical Review Letters* **93**, 077207 (2004)


PERZLMAIER, K.; BUESS, M.; BACK, C.H.; DEMIDOV, V.E.; HILLEBRANDS, B.; AND DEMOKRITOV, S.O.:
Spin-Wave Eigenmodes of Permalloy Squares with a Closure Domain Structure
*Physical Review Letters* **94**, 057202 (2005)


BUESS, M.; RAABE, J.; PERZLMAIER, K.; BACK, C.H.; AND QUITMANN, C.:
Interaction of Magnetostatic Excitations with 90 Domain Walls in Micrometer-Sized Permalloy Squares
*Physical Review B* **74**, 100404(R) (2006)


NEUDECKER, I.; KLÄUI, M.; PERZLMAIER, K.; BACKES, D.; HEYDERMAN, L.J.; VAZ, C.A.F.; BLAND, J.A.C.; RÜDIGER, U.; AND BACK, C.H.:
Spatially Resolved Dynamic Eigenmode Spectrum of Co Rings
*Physical Review B* **73**, 134426 (2006)


NEUDECKER, I.; PERZLMAIER, K.; HOFFMANN, F.; WOLTERSDORF, G.; BUESS, M.; WEISS, D.; AND BACK, C.H.:
Modal Spectrum of Permalloy Disks Excited by In-Plane Magnetic Fields
*Physical Review Letters* **96**, 057207 (2006)

BACK, C.H.; PERZLMAIER, K.; AND BUESS, M.:
Dynamic Aspects of Magnetism
in:
BEAUREPAIRE, E.; BULOU, H.; SCHEURER, F.; KAPPLER, J.-P. (EDS.):
Magnetism a Synchrotron Radiation Approach
Springer Verlag, Berlin (2006)


MARTIN, T.; BELMEGUENAI, M.; MAIER, M.; PERZLMAIER, K.; AND BAYREUTHER, G.:
Pulsed Inductive Measurement of Ultrafast Magnetization Dynamics in Interlayer Exchange Coupled NiFe/Ru/NiFe Films
*Journal of Applied Physics* **101**, 09C101 (2007)


HOFFMANN, F.; WOLTERSDORF, G.; PERZLMAIER, K.; SLAVIN, A.N.; TIBERKE-VICH, V.S.; BISCHOF, A.; WEISS, D.; AND BACK, C.H.:
Mode Degeneracy due to Vortex Core Removal in Magnetic Disks
*Physical Review B* **76**, 014416 (2007)


BAILLEUIL, M.; HÖLLINGER, R.; PERZLMAIER, K.; AND FERMON, C.:
Microwave Spectrum of Square Permalloy Dots: Multidomain State
*Physical Review B* **76**, 224401 (2007)


PERZLMAIER, K.; WOLTERSDORF, G.; AND BACK, C.H.:
Propagation and Interference of Spin Waves in Ferromagnetic Thin Films
*Physical Review B* **77**, 054425 (2008)